A Few Good Memories

by Soumen N. Ghosh

The contents of this work, including, but not limited to, the accuracy of events, people, and places depicted; opinions expressed; permission to use previously published materials included; and any advice given or actions advocated are solely the responsibility of the author, who assumes all liability for said work and indemnifies the publisher against any claims stemming from publication of the work.

Dorrance Publishing Co
585 Alpha Drive
Suite 103
Pittsburgh, PA 15238
Visit our website at *www.dorrancebookstore.com*

ISBN: 978-1-4809-4545-6
eISBN: 978-1-4809-4522-7

A Few Good Memories

The Big Tusker

Chapter 1

Getting Ready

"Get up, get up, Babu, wake up. We have to go now; the driver is waiting in the truck. He will show you the Big Tusker, the elephant that I told you about. I saw him the other day while the workers were loading the logs on the back of the truck. It's a big elephant almost 12 feet tall, must be 8 tons. I have never seen that big of an elephant." Biren, Babu's father, now gets impatient at the tardiness of this little five-year old boy. "Oh, Babu's Ma, get him ready, it is almost five. We must leave now. We must go to Musabani by 5:30 A.M. and it will take another hour to go to the peak, then he can see the elephant."

Babu, Bubu, Bablu, or Bubla are all the nicknames of Samar. Samar Sen, the famous architect who designs multistoried buildings or what they call skyscrapers of the world. Last year, Samar designed 21 skyscrapers,

50 stories or taller. His office has 34 staff, including 15 junior architects, designers, planners, etc. Samar's latest project was in Bombay, an old Jewish building that was almost dilapidated and about to be condemned. Alex Rogers, a Wall Street broker and philanthropist from New York, has a connection to this building. His great-great-grandfather was a Mennonite who moved from Jerusalem to Bombay for starting his fortune in the spice business. And man, he made a real fortune. Most of the family later migrated back to either Israel and some moved to Miami and a few others to New York. Jim, Samar's friend and one-time classmate at the graduate school at Berkley, told him one evening when they were talking to Zubin at the back of the stage after Philharmonic brought back Zubin for a three-day gala event in celebration of Lenny's 75th birthday.

"Sam," Samar's nickname in the US, "you must meet Alex. Remember, I told you about the Jewish temple that he wants to renovate and rebuild in Bombay? He wants to meet you. Actually, he is right over there. Let's go and talk to him over a cup of coffee, what you say? Then you can decide if you want to take up the project. Okay?"

"Hey, Alex, meet Sam, the famous architect."

"Yes, I know, then who doesn't know about Sam, especially after his project in Central Park?"

Yes, Samar did an awesome design at the north end of Central Park.

"Jim, let's go over to the 5th and 42nd, I know a very good and quiet coffee shop, they are open till 2:00 A.M., and it is only 10:45," Alex asks.

Samar is not in a mood to have a late-night coffee, and Sue is not well, her right knee has been bothering her for a while and he must go before too late. "Well, perhaps another day," Samar grumbles.

"No, no, I insist. It will take only 15 minutes. How often can we have this opportunity?"

"Well, then."

The three walk down Broadway, to 44th, and then cross over a few blocks to the 5th and there is the obscure coffee shop—"Jane's Coffee." Very simple. Samar knows New York like the back of his palm. He moved from upper part of the town to the Manhattan central area about four years ago. They bought a 6th-floor apartment on Lexington and 37th Street. But never came to this side of the 42nd and definitely never saw Jane's Coffee. As he enters the shop he looks back at the right side of the street, something doesn't seem right. His analytical architectural mind tells him that something is very odd about this store.

"Sam, how would you like your coffee? Black? Or with cream and sugar?" Jim asks.

"No, black only."

"Well, Alex, tell him about your project," Jim points to Alex.

"What project?"

"That one! That your great-grandfather tried to finish?" Jim quips.

"Yes, yes, Sam, I want you to take up the project and show the world what my great-grandfather meant to the mankind, and I also want to show your brilliance to the rest of the world, not just glass and steel work that made you so famous. I want to show how an Indian vision and the Jewish money can do. Will you, will you take it up, my friend?"

"Okay, let me think for a day, and I will call you tomorrow by 9:00 P.M. I must get going now, Sue is waiting for me."

"Bye."

Samar met Alex the next day as planned. Outside New York Philharmonic, around 10:30 P.M., these three were bidding goodbyes to each other after the stupendous concert conducted by Zubin Mehta. Samar was casually saying, "I just talked to Zubin on Monday; he wants me to visit him in LA, where his 90-year-old dad lives and wants a different kind of amphitheater in Bombay for their youth orchestra." That was four years ago today.

The nine-story building or one may call an edifice is officially only nine stories, but feels like at least 20 stories. But that's another story. Today is the inauguration of that building, they call it "Man-of-the-Isle." The Man-of-the-Isle is completely lit in blue and white lights, just like a tall symbol ready to direct the ships in the right direction. But Samar is not there, at least not mentally.

A gala event has been organized at a posh hotel in Bombay for the opening of the "Man-of-the-Isle" Jewish temple. Well-dressed men and women are milling around in front of the front desk, a huge picture of the "Man-of-the-Isle" is placed and a small insert of Samar's picture as the architect. Samar is quite embarrassed to see his picture, but he takes it in his stride and is shaking hands with guests. Suddenly, his cell phone rings, he looks very disturbed (frown on his face) and steps aside and takes the call, looking very distraught. The opening ceremony is about to start in 15 minutes. He hurriedly approaches the head Rabbi, Mr. Joseph Silverman, and murmurs a few words in his ears. The Rabbi puts his right hand on Samar's back and Samar leaves the scene hurriedly.

He just received a call from his mother. "Babai is not well, come home quickly if you can."

He was already planning to go the next day after the inauguration of the "Man-of-the-Isle" temple on the far west side of Bombay, which he designed about three years back. Last night was the gala opening and tonight a big "black-tie-only" party for an exclusive list of guests, but he can't attend. He must go…must go now.

Jet Airways flight to Calcutta is about to take off from Bombay, the time is around 6:45 P.M. Samar is in the business class seat, settling down with his briefcase and a backpack. The plane is waiting in the tarmac for takeoff. Cabin lights have been dimmed off, only a reading light is on next to his seat, sort of above the shoulder comfortably raised for direct focus on the subject at hand. He is looking at today's *Wall Street Journal*, which the hostess put nicely folded next to his seat. He is looking at the front page but not reading; his gaze is somewhere else.

Something is constantly buzzing in his head—*Wait till I show you the Big Tusker, a huge male elephant with beautiful six-feet-long white tusks, gorgeous, Big Tusker… big…big….*

A gentle touch wakes him up. "Sir, please put on your seatbelt, we are about to take off."

"Oh, sorry, thank you."

Only a return smile comes from the stewardess. "Would you like a drink, sir?" the stewardess asks.

"Yes, scotch, single malt, if you have it, and straight on the rocks. Oh, and please don't wake me up if I fall asleep, and I don't need any food, thank you."

Again, a gentle smile back from the young and smiling.

Chapter 2

Down the Memory Lane

The lorry (old British term for trucks) was painted in green and red during last month's Viswakarma Puja, green on the outside, that is the wooden panels for all three sides, making it look like a box to hold the cut-logs, and bright red on all three sides of the driver's cabin except the front. The front of the lorry, the bonnet, was painted part green and part red, giving it a weird look. Babu did not like the colors, he wanted yellow on the front and black on the top looking like a tiger, but Biren said no, that will scare the animals in the jungle. Babu's little brain did not quite understand why yellow will scare the animals off but not this ugly red and green combination. The little boy somehow accepted his father's decision, what could he have done anyway? Fathers are there to make such decisions. He did not complain much again.

Samar vaguely remembers the factory shade, more like an empty, relatively large structure with tin rooftop and a little elevated murram based groundcover. On one side of it (about a 10 foot by 10 foot square has been set aside), where the Vishwa Karma deity is being set up for its annual worship and invocation. The lorry (truck) is parked a little back about 50 feet from the auspicious place, and about to be dressed up since this vehicle is the main source for Vishwakarma's and for the family's daily livelihood.

Babai (Samar's dad) has been very busy since morning for the puja, which starts at noon and will continue till 2:00 P.M., after that everyone (all the workers and the invited guests, about 20 of them) will partake the prasad and then usual feast of goatmeat curry, rice, etc., will take place. This goat meat curry is a special dish, and Babai himself cooks it. He has been doing this for a few years since he has started his logging business. Now that he has no other business partner, he must look after everything to the best of his abilities.

A couple of people are painting the lorry with just a hand brush and opened cans of paints. Back in those days there was no machine painting, and that too not in Ghatshila.

"I don't like the color, babai," little babu (Samar) tells his father, "'specially the front. Why did they paint it in green and yellow? And why the back is all black and look at the sides, all red. Yak, yak. Ekdom baje, ekdom bhalo lagche na. I am going to tell Ma, it is very bad, I don't like it at all." Babu looks very upset with colors that they painted the lorry with.

"Come here, Babu, let me tell you why they have colored the lorry like that."

Babu goes to his father and Biren places him next to him and both look at the lorry being finished and decorated.

"Look, Babu, when you are in the jungle going up slowly, lots of animals stare at you. If the color is too bright, they get very curious, monkeys, black bears, and even cheetahs may jump on to the lorry, and trust me, that happened to us once. Not only that, elephants don't like bright colors, they get scared, and you don't want scared animals. That is one reason I asked them to paint it like that in multiple colors, so the animals get confused and don't come out of their hideouts and we do our job. Son, don't you want your father to come home unhurt every day, and more importantly, I don't want Magan Singh getting scared and driving us into the gorges. Do you now understand?"

"Yes, Babai, but I don't like it."

"Okay, trust me, you will get used to it. Now go inside and put on your new clothes and call Ma and your little sister. We are about to start the pujo."

"Are you sleeping, son? We are crossing the Mussabani Bridge, we should be in the forest within 20 minutes, keep your eyes open, you will see lots of little animals as we start climbing." Biren opens his flask and pours a little more hot tea into his cup, and asks his son, "Do you want some tea?"

Little Babu is not used to tea at all, and obviously he does not like waking up, with a half-asleep voice and disgust, he says, "No."

Biren goes right back at his son: "It is all right, have some, this will keep you awake, and you will not fall asleep.

We must stay awake to see the Big Tusker, don't we?"

The lorry is crossing the bridge on Subarnarekha and starts ascending. Babu can feel the climb, speed slow, and the lorry starts to emulate a sound that resembles an old man carrying a heavy load on his back, grunting, breathing heavy, and complaining about his joint pain, just not happy. But Biren whistles a Santali tune; he plays flute, but he is just whistling, Babu is sleeping with most of his body leaned against his father's.

Where Biren has leased about 120 acres for logging is farther up and almost at the top of the hills on the southeastern side of the Dungri No. 9. Mussabani is all about hills. These are all parts of the Dalma range of hills and have large deposits of minerals. Mussabani and the hills surrounding it are rich with copper ores. In between those hills there are large tracts of forests with all varieties of tress. These forests are leased annually through a bidding process for logging of timbers. Biren bade last year (1958) and won the bid for a mere 300 rupees per acre of forest land. He may not clear-cut, but is allowed to cut any tree that is 16 feet or taller. Biren is quite happy, he is hoping to make a very handsome profit that year and perhaps continuing to log for another few years before he goes back to Calcutta and opens a small manufacturing facility. He is very good with designing small machine tools.

Biren left Calcutta with his young wife and a two-year-old son. He came to Ghatshila as a superintendent engineer for the Dhalbumgar King's mine factory. This factory makes mine lamps. These lamps will glow in dark with very little oil and also will detect the presence of any methane gas that is so prevalent in all these mines in this region.

King Ram Singh is an educated man, he is a lawyer, a barrister by training, and is residing and practicing in Ranchi, a relatively bigger town in those days in that part of Bihar. However, he has this passion for producing these lamps that are so important for the mine communities. This whole area of about 300 square miles is full of a wide variety of mines, and all of them need mine-lamps. Ram Singh himself hired Biren when he was staying at the Great Eastern Hotel in Calcutta after a hearing at the Calcutta High Court. Biren was looking for a job and he was very active in college politics. Biren never kept a steady job. Even after marrying his college sweetheart and comrade in politics in spite of all odds and uncompromising parental disagreement, he did not compromise. He took various odd jobs while his wife was preparing for the final examination for the bachelor's degree with English honors.

Prativa, Biren's wife, is very bright, always standing first in her class, and is a pride of her parents. Her father wants her to be a professor of English literature. Prativa is comfortably progressing to fulfill that goal of her father as well as her own. Then, while in the second year of college all her plans and focus got diverted to Biren. Biren is a final-year student and a leader, his speeches on Marxian Philosophy and a clear rendition of dialectical materialism and the inherent contradiction in mankind in terms of ego and superego got Prativa moved.

After that lecture, she could not resist meeting Biren, and the rest is history. After their marriage, Biren and Prativa stayed in Biren's parents' home for about a month. Biren found a one-bedroom flat (apartment) in north Calcutta and moved there. Prativa could not stay in that place for more than two months, for she was already three months pregnant with Samar, and her college was also too far. Biren had to move back to his father's house again.

Samar was born on a regular Friday morning at the Bangur Hospital at Tollygunge. The year was 1954. The new independent country is less than seven years old, just learning to walk without the Masters (Britishers) dictating terms. The name Bangur came from the Pali-Merwar family of Marwaris. A trading class, however, dedicated to education, health, and infrastructure. The Bangur Hospital was established in Calcutta during early 20th century under Lord Curzon.

Indian parents do not name their offspring before birth. It is considered unwise, and unlucky for the child, perhaps because death rate back in those days were pretty high, and because of the mortality factor it was considered "Aashubha," or unlucky to name the child prior to his or her birth. Samar's birth certificate reflects that, when Samar got his US citizenship, he had to produce his original birth certificate and there was no name, it took him quite an explaining to do before the authorities and the Americans obliged.

But what will they call him? Biren embraced the little one and told Prativa, "Look at my Babu, he is so fair in complexion, just like you." Biren was rather on the darker side, and Prativa was very fair complexioned. Prativa pronounced her son as Bubu, and held him high. They came to Biren's father's house with not much fanfare. But when relatives started to peek into the little room where the mother and son were staying, everyone wowed. Prativa's parents didn't come running. It took them another day to come and see their grandson. When they saw him, they were overjoyed mainly because Babu or Bubu did not look like the Kayasthas, he looked more like the Brahmins with that fair skin and even more.

"Sir, we will be landing in 30 minutes, do you need anything else?"

Samar wants to say, "Just go away, let me be in my dreams," he just yawns, and the cabin attendant understands.

The lorry is making even more noise as it starts climbing steeper hills. At some point Babu thinks that it will stop, but doesn't say anything, he is still very sleepy. Biren is saying something to the driver, Babu can hardly hear, even if he heard, he would not understand a word because those are some Bihari dialects that were only spoken in that part of Bihar. Even though the main language is Hindi, but every district in Bihar, for that matter all over India has its own little dialect. Biren is quite conversant with those that got him going in that desolate but attractive place. He is always very adventurous and this challenging task of logging in a far land from home, Calcutta, does not bother him. He is about 30, and nothing can tame him down.

Suddenly the lorry stops with a jerk, and Babu wonders why, his half-asleep eyes are not ready to roll around and search for wonders of the forests.

BIREN SOUNDS PRETTY EXCITED:

"Babu, look, look at those peacocks crossing the road." As per Biren there is a whole family of peacocks with their full fluffy feathers and one male peacock is slowly crossing the road with his feathers fully spread.

"Look, Babu, look, look at those Mayurs."

Babu is still sleepy.

"Where is my Big Tusker?" is his reply.

"It will come soon, just look at the peacock."

By the time Babu peaks from the window from Biren's lap, the Peacocks are gone.

"Magan Singh, let's go, it is about 5:45 A.M., pretty soon the sun will start beaming, we must reach mile marker 7 before the elephants cross over to drink water."

Magan, the lorry driver, starts the engine back. The engine grunts, old and tired, does not want to move anymore at that early hour, perhaps a few more hours of sleep is needed.

The lorry starts rolling again. Winding up and up, and Babu still sleeping on his father's lap.

He is feeling a slow motion and a steady noise that he never heard. He wakes up hearing Biren talking to Magan Singh.

"Singhji (honoring Magan Singh, the driver), what do you say, we take the lorry to Jamshedpur this Sunday? Bohoot awaj karta hai (making too much noise). I think it needs some valve job, even may be some compression problem."

Biren is not an auto mechanic by training, but he has a thing for all things mechanical. His intuition about engines, or for that matter most things, is near perfect.

"Hnah Saab, Etwar ko ja sakte hai, me re kheyal se gasket, valve vi check karna hai." ("Yes Sir, we can go to Jamshedpur on Sunday and will check things out. I think we need to check the valves and the gasket.")

Magan Singh is a pure Bihari, he does not know how to read or write, but knows driving very well. He does not know much of anything inside the engine. He only knows how to put water on the battery shell, and start the engine without pulling the choke or starter. Back then there used to be chokes to jumpstart the engines. The lorry is ten years old. Biren cannot afford a new lorry, his partner from his quarry business was selling this lorry when he and Mr. Chaudhury parted company and diluted Biren's share of the quarry business. Biren settled for only eight thousand rupees. Since then he spent more than double that amount fixing the lorry. Prativa always complained, saying, "Why did you buy this junk, and spending more and more every day? Women have better sense of these things." Biren did not say anything in return.

The lorry was making its way up slowly but surely. It will take another ten minutes or so, about two miles before there will be a stream, and they must cross that stream and go another half a mile to approach Biren's leased area. This part of the uphill is full of surprises. Biren always found something new. Every night when he returned from his work he used to tell Prativa something new he did find that day, whether it could be a new wildflower, a new tree with different kind of smell, or a different type of a monkey, or even a snake that is unique to the area. He was only 29 years old, new things come automatically at that age. He was not in a mood to find anything new today, only after one thing—the Big Tusker. He promised his young son that he will show him the big one. Let's see if Lady Luck will smile on him!

The lorry starts to roar, it is going in first gear since there is loose gravel, small boulders, and the cross-current of the stream. Little Babu is in deep sleep, his little head is leaning against his father's arm, too small to reach Biren's shoulder. Biren puts his scarf, folded, and makes like a little pillow and puts that between Babu's neck and his arm so that the little one can have some comfort. But the agonizing slow and jerky motion of the lorry makes him constantly adjust the cushion. It does not bother Babu at all, he is in deep sleep and dreaming.

"Magan Singh, jara rokhna (stop for a second), pissab korna hai" (I have to take a leak). "Saab, idhar nehi rokhna, porsu bhalu nikle tha idhar, thora age ja ke rokhunge" (a black bear was seen here, sir, will stop at a place little further). Thik Hai, Biren answers.

Magan stops after five minutes, Biren gets down and relieves himself, Babu is still sleeping, and does not wake up. It is about 5:50 A.M., the sky starts to turn orange and some patches of clouds are fast passing, Biren is dosing off a little; he didn't get much sleep last night, had to work on the accounts till midnight and then woke up at 4:00 A.M. He was dreaming—a full load of Saal timber, and some Segun (teak wood) will fetch him a good bit of money from the Howrah Mill, where he has been supplying these logs since past four months. A full truckload will be about 12,000 rupees. Out of which he will pay back the loan for the truck, that is about 8,000, and the rest, he will invest on a small plot of land, about two bighas, next to Mr. Chaudhury, well, a little further from Chaudhurys' but closer to the river. Prativa wants her own house. He is quite engrossed with his happy dream and floating further on to pleasantries.

A big jolt, rather a hard break with a grinding noise, tears his dream apart. "Keya Hua Magan Singh? What happened, Magan Singh?" "Saab chup, chup (don't yell) look up there, Saab, Hathis nikle hai, sube ka pani pine ke liye; elephants are just coming down for their first drink of water of the day, be careful. Babu saab ko uthaye na" (Why don't you wake up the little master). Hanh, Hanh; Babu, Babu, get up, look, look outside, Hathis are coming."

Babu is startled, he is still very sleepy, he peeks through the window, nothing can he see, he is still too little and cannot reach the viewing part of the window.

"I can't see any Hathi, Babai."

"Magan Singh, Babu Ko apke gaddi me utaho (Magan Singh, please put him on your lap and let him see from the window on your side)."

Chapter 3

The Appearance

Do the elephants kick their left foot out first? Or the right foot? Samar once attended a show in Santiago, Chile, where a puppet show was performed by a French cartoonist, Oliver Seigrid. That was almost 15 years back, when Samar was attending a UNESCO conference on "sustainable architecture in the 21st century." Oliver was also a ventriloquist, but unlike others in this profession, Oliver was actually physically mimicking elephant's walks—left hind leg goes up a little, no stomping, gently touches the ground with the paws, and like a baton rising up and down in the hand of a maestro, lifts the right front leg and the paws touch down. Then the right hind leg and the left front leg, with a pause of about a second touch the ground. Oliver was in an elephant's suit. The trunk was too big for the body and did not match at all, that was the instant reaction Samar had, and his mind's eyes could not concentrate on the rest of the walk.

"Babu, look, look at the mother Haathi (elephant)," Biren almost screamed.

"Saab ayestha (speak softly, please), Uska bara kaan hotha hai, Un ne Nehi Ayega. Elephants have long ears, they can listen from afar, she will not come out," Madan Singh quips.

"I can't see anything."

"Oh, ho! Wait, you come to the right side, Magan Singh, please put Babu on your lap and let him see through the window from your side. Can you see now?"

"Yes."

"Chup, Chup (please keep quiet). There they come!"

The mother Hathi was carefully guarding her calves, four of them. First she pushes one out, then there were two others, she gently lashes their buttocks with her trunk, the three little calves slowly walk, as if the mother was pushing them from their beds. The fourth one was not coming out, actually she was clinging to the mother's tummy. Mom steps back a little, stretching her left hind leg and then allows her tail to swing back and gently touches the little one. She sort of springs out from the bottom. All of them, the mother and the four little calves, start walking slowly crossing the metal road.

"Where is the Big Tusker?" Babu asks his father.

Biren was watching the whole family crossing the road. He was also getting impatient for not having been able to keep his promise to his young son—showing Babu the Big Tusker. He asks, "Magan, where is the Marad? The male elephant?"

Magan just says, "Oh aaj Nehi ayega" (he is not coming today), as if he is too tired and lethargic."

Babu understands his Hindi, and looks at Magan Singh, and says in perfect Hindi,

"Jaroor Ayega."

Biren looks at his son and got surprised, he never heard his son speaking in Hindi. "Okay, Babu, aap dekhte jao (just keep watching). Magan was about to start his engine, suddenly, he heard the murmur and the little chirping of the birds and the subtle sound bites of stomping and braking of twigs and little branches of lots of trees.

HERE COMES THE BIG TUSKER!

"Oh, Babu, watch him, see how big are the two tusks, see how nicely shaped and pure white, this is the Master of this jungle, my Big Tusker. He came out just to show you and help me keep my promise to you."

Biren was so excited to show his son what he has promised him for so long. It is like a vindication and pride and also some sort of self-respect. After all, he is the father, and father must keep his words to his son. He was content.

Babu was all wide awake and did not pay any attention to what his father was saying. He was in a dreamland and quite mesmerized by the sheer mammothness of the Big Tusker. The big one was totally oblivious to the crowd, even though all were in that truck. He is quite used to it and knew that nothing wrong these humans can do to his family. He was like the big shield protecting the family gently moving toward the stream for some early-morning sprinkle and fresh water.

Babu's eyelids were not closing, all wide apart and consumed by that heavenly walk at a very slow pace: left hind leg goes up a little, no stomping, gently touches the ground with the paws, and like a baton rising up and down in the hand of a maestro, lifts the right front leg and the paws touch down. Then the right hind leg and the left front leg, with a pause of about a second touch the ground, all in unison, Again and again all five, and the Big Tusker marching on.

Ugly Leg Cramp

Chapter 1
The Match Begins

The match (tennis) was just about to start. It was a quarterfinal. The winner will have to play again today and will move on to the final for a show-down on Sunday. Shoumit (that's how he pronounced his name, even though the correct spelling is Soumit) looked at his father for an approving nod.

"Okay, my boy, go get 'em. Remember to toss the ball high and try not to lose the first serve on to the net."

Suman, Soumit's dad, repeatedly uttered these words as they were driving to the far end of El Paso, American Eagle High's tennis complex. The first match was scheduled for 11:0 A.M., and it takes about an hour down the I-10 East from Las Cruces to reach this far. They started around 9:0 A.M. so that Soumit will have enough time to practice with his teammates before his one-on-one. But leaving on time was a far cry. He was not ready, still playing Nintendo with his little brother. His mother had to almost push him out of the house around 9:15. "Go NOW," was the ultimatum from Neela. Finally, he gathered his tennis bag, extra pairs of socks. Baba (Suman) already took his water bottle, couple of bananas and a box containing chicken-salad sandwiches for lunch. Neela woke up early for her son and husband and packed a solid lunch.

"Baba, what if I win the first match and go to the Semi but can't win the Semi? Will you be still proud of me?"

"Of course, I will be proud of you regardless. I am always proud of you, but you must win that too. You must always try harder and harder. Win or lose is not in your hand, but you must and always try harder.

"Try not to make any silly mistake. Try not to have any double faults. Toss the ball little higher and as Sally (the coach) told you repeatedly, do not try to volley from so far behind, let the ball drop and then use a double-handed backhand or hit it down the line with your forehand; you have a very good forehand, use it to the best of your ability."

Such was the advice at the moment from Suman. Soumit started adjusting his cap. He always wore caps. He was wearing a red cap with a gold star. First he put it backwards, didn't like it and then readjusted it properly as he was about to do a practice serve against Ben, another Sally team member. This was a regional tournament. All local (within 100-mile radius) high school juniors and seniors, including Ciudad Juarez High School boys, were eligible to qualify. Initially, the first three rounds were done locally; his was in Las Cruces. Eight high schools competed, and only one team of two boys between the ages of 16-18 made to the third round. Yesterday was the fourth-round match and Soumit had an easy win. Today is the quarterfinal followed by the Semi. The fixture was posted outside the tennis complex on a big board. Soumit was in the winner's bracket and will play against Nathan Jones of East Side High of El Paso at 11:15 A.M.

Suman never could figure out how these boys decide who will serve first. Back in his days, the umpire would come up to the net and toss a coin high up and ask one of the players to call head or tail. Things are different now. Soumit puts his racket head down and gives it a spin. When the racket stopped spinning, and lay flat on the ground, the head was in the opponent's direction. That meant that his opponent won the toss and has the privilege to either serve first or chose the side. Nathan chose to serve. Suman looked at his son and waved him with a "V" sign. He was sitting next to another parent of the Sally-team. Janice, mother of Jake and Crisse, also came to support Sally-team even though Jake, her son, lost yesterday and did not make this quarterfinal round. She smiled at Suman and said, "We love Soumit and I am sure he will win."

"Thanks," just a short reply. Suman was deeply engrossed staring at his son.

Chapter 2

Set One

Nathan had an Ace, his very first serve was a killer. Soumit looked firmly at the spot where Nathan's serve hit the ground and started fixing the strings. This is a routine thing that all professionals do these days, they lose a point and then start working on their strings, as if it was the racket's fault.

Second serve, another ace. Umpire calls the point: "Thirty – Love."

"Come on, Soumit, cheer up, cheer up."

Soumit did not return the glance.

Nathan serving his third serve. Oh my, yet another Ace. The crowd goes loud and started clapping and screaming, "Go, Nathan, go...."

"Forty – Love."

Nathan serving for the game. He serves outside the second line.

"Fault."

Goes for his second serve. Nice serve right in the middle and little to the right of where Soumit was positioned. Soumit returns with a cross-court down the line forehand, Nathan runs to return the ball, the ball flies over the net but way over the line. Soumit wins the point.

"Forty - Fifteen."

"Attaboy, keep it up," Suman cheers.

Nathan gets ready to serve, straight into the net.

Umpire calls, second serve; oh, again straight on to the net. Lost that point also:

"Forty - Thirty."

Nathan looked a little disturbed but gathers his thoughts and goes back to the serving line.

Nice serve, little on the left side where Soumit was waiting for it to come. He waits for the ball to come waist high and picks it up with a doublehanded return, Nathan hits hard and this time was successful to keep it on the court, Soumit hits a strong forehand down the line and before even Nathan could reach the spot the ball landed second time and Soumit wins the point.

"Deuce. Forty all."

Nathan looked a little puzzled, he gathered himself and tosses the ball high and serve. Another Ace!

"Advantage, Nathan." Soumit looks at the spot and looks at the crowd, went back to his position to receive the inevitable.

Nathan tosses the ball, rather low, hits it hard, the ball clips the top of the net and aced Soumit.

"Late." Nathan gets ready for his serve again, this time he tries to hit a top-spin, the ball came short and a fault.

"Second serve."

Nathan serves hard, "Out."

"Deuce."

This time the serve was right on the center swerving to the right of Soumit, he had a perfect opportunity to return big with his forehand and down the line. Soumit exactly did that, it was beyond Nathan's reach.

"Advantage. Soumit."

Nathan was getting frustrated, the crowd tried to cheer him up with chanting, "Nathan, Nathan, Nathan."

Oh, no! Nathan serves straight on to the net; he slams his racket on to the net; goes back mumbling. Serves hard, but this time it was way too long.

"One – Zero, Soumit," umpire announces in a loud and firm voice.

"Come on, Soumit, come on now." Many voices said the same thing from the Sally camp. Suman just smiled; he was beginning to feel proud of his son. He looked far, Soumit was getting ready to serve his first for this match. Suman heard himself saying, "Don't serve on to the net, toss the ball high."

Soumit serves a perfect serve, little to the left of Nathan.

Nathan returns, almost missed the net, somehow the ball dropped on Soumit's side and bounced high, it was an easy finish for Soumit with a smash to Nathan's discomfort.

"Fifteen – Love."

Soumit served an Ace. "Thirty – Love."

Suman looked at his son but didn't get a return look back.

The next three points were relatively easy and Soumit now at 2 – 0.

The third game went in Nathan's favor. Soumit couldn't return any of Nathan's Aces. He hit perfect four Aces in a row. El Paso Crowd goes gaga for Nathan. He regained his confidence. Looked strong. It is Soumit's turn now to show some class.

First serve, straight on to the net. Tossed the ball low, a chronic problem for him.

Suman shows frustration. Ahh! Ahh!

"Second serve." Soumit hits a top-spin, relatively strong serve, Nathan tries to go against the spin and hits straight outside the second court. Fifteen – Love.

Soumit adjusts his cap, hits an Ace. Thirty – love. Sally team cheers, "Soumit, Soumit, Soumit."

He hits a long serve right on the baseline, Nathan returns, and it was a perfect volley for Soumit.

"Ya, ya," he pumps his fist.

"Forty – Love." He then serves an Ace.

"Three - One." Soumit.

The next three games were rather boring. Since that double fault, Nathan could not regain his confidence, he had two more double faults and lost the fourth game forty –love.

The fifth game went straight in Soumit's favor with two consecutive strong Aces, and the last game of the set was again a disaster for Nathan.

Soumit wins the set 6-1.

Chapter 3

Set Two

Straight on to the net. "Second serve." This time it was a beautiful serve, coming straight to Nathan's body, before Nathan could move to his left and attempted a forehand it crossed the line and landed on the far end fence.

"Fifteen – Love."

First game was very easy, Nathan could not gather himself at all. The next three serves were either returned on to the net or out of bounce.

Second game was a mere repeat of the first. This time it was Nathan who double-faulted twice and could not return Soumit's strong volleys.

"Two – Zero," Soumit serves.

First serve was an Ace, Soumit pumps his fist, "Yah, yah."

"Fifteen – Love."

Next three points were just routine.

Soumit is up three zero.

Nathan looks pale. El Paso boys cheering in unison—"Nathan, Nathan, come on, Nathan, don't give up, don't give up, do your Aces."

Nathan just did that. Three beautiful Aces.

"Forty – Love."

Soumit looks at the bleachers, met Suman's eyes and gets ready to take the final serve for this game from Nathan.

Oh, my. Straight on to the net.

"Second serve."

"No, way," one of Nathan's friends lamented. The serve was out of bounce.

"Forty – fifteen."

Nathan goes back hits the ball again on to the net. He got furious, slams his racket onto the concrete so hard that it got bent at the top. He runs to the sideline and opens his tennis bag and pulls another racket with a brand-new strings. He tests the strength of the new strings by pushing it with his palm, felt good and goes back to the serving line.

He tosses the ball high and bends over and serves an Ace.

"Fault," the umpire shouts, foot fault. Nathan gives the umpire a very angry look.

"Second serve."

Oh, NO! Straight on to the net.

"Forty – Thirty."

The next two points were just giveaways.

"Four - Zero, Soumit."

The last two games were very ordinary and pretty disaster for Nathan. His anger got the better of him, and he could not return any of Soumit's top-spins or volleys.

Soumit wins, straight set. 6-1; 6-0.

Now, it is about 1:10 P.M., the Semi will start around 3:00 P.M. By the time Soumit gets to play the winner of the other quarter final, it will be close to 4:0 P.M.

"Dad, I am hungry, let's go for lunch."

Well, your mom sent our lunch, let's eat that. "

"No, I want to go to Burger King with my group."

"Okay."

Chapter 4

The Semi

After a grilled-chicken sandwich and a large Coke Soumit wanted a quick power nap. He puts his Sony Walkman and earphone and stretches his long legs up on to the dashboard of his dad's green pickup Nissan. Suman didn't bring the Previa this time. Neela and Krisch have to go for Krisch's cello lesson at the music school.

The ride back from Burger King to the tennis center is only three miles. Soumit has already fallen asleep, his deep breathings gently rumbles in Suman's ears. "What if I don't win the Semi? Baba? Will you be still proud of me??"

He looks at his son's face, very serene, calm, and oblivious to the upcoming match that he so desperately wants to win.

Is winning everything? Who says it is? Then again winning only counts, winners go to places, and losers just dream. Such is the proverbial belief in the system. Suman is a winner in some cases, but losers in many other fields. But it is the winning that has carried him this far, got him his Ph.D., a professorship in a reputable university, all that. On the other hand, had he not won; life would go on. Back in Calcutta, he was fine as a ranked-officer, pushing millions of odds and making his strides worth.

"Baba, we are here." Soumit quips as the truck stops at the stoplight just before the tennis center.

"Oh, I see. I did not notice. So, how do you feel, my son?"

"I am great, just a little tired."

"Well, you still have 45 minutes before you check in. Why don't you take a little more nap?"

"Okay."

Soumit closed his eyes. "Can't sleep."

"Never mind, just close your eyes and do not think about tennis, the semi, or anything. Just close your eyes, son."

That was the simple advice Suman gave him, as if Soumit will immediately listen to it and all the fluttering in his mind just miraculously disappear and he will be calm and composed and ready to take on the inevitable. Soumit sat up on his seat and put on his famous red cap, backwards, adjusted it a little.

"I am getting off the truck and going to be with my teammates over there. See you after the match, Baba."

"Good luck, and don't forget to take your water bottle, fill it with some more ice, they have an ice vending machine, next to the restrooms."

"I know, I will do that, don't worry."

Suman wandered around for a few minutes, went to the men's room, and got back to the same spot on the bleachers.

"Hi." This time it was Sally, Soumit's team manager-cum-coach, and owner of the tennis club that got him up to this point.

"Don't worry, Soumit will play an awesome match and win this tournament. I have faith in him, and we are all counting on him."

"Thanks, Sally, thank you for your confidence."

The match is about to begin in ten minutes.

The match lasted for 2 hours and 51 minutes. First four games, Vasquez did not make a single mistake. Soumit was down 4-0.

Caesar Vasquez is a stocky guy, 5 feet 6 inches, hails from Ciudad Juarez. He wore his cap backwards and does not show any emotions. He was a State (Chihuahua, Mexico) champion and was the top favorite for this open regional championship for all-boys' senior (high school) champion trophy. Everywhere he goes, he carries a dozen or so of his fellow classmates and all his family (about 20 of them) members. They go "Hail Caesar" every time he wins a shot.

The fifth game was just about to begin, Soumit fixes his cap and toss the ball high enough to serve an Ace. "Go, Soumit, go, Soumit," the entire Sally camp shouted at the top of their lungs. Suman kept totally quiet, not even a gaze back to his son.

Soumit won the next three games. The score was 4-3. Crowd was going very noisy. Caesar served his first serve straight on to the net.

"Second serve," the umpire announces.

Caesar tosses the ball and hit very hard, it aced Soumit, but wait, umpire calls, "Foot-Fault."

"Oh, no, Love – Fifteen."

Vasquez casts a nasty look at the umpire but decided not to say a word. Hits a top-spin down the line and a winner. 15-15. He wins next three games by sheer power and aces. Score is now 1-0 Vasquez.

The second set was a breeze for Soumit, it started like this:

"Double fault," Mr. Chavez, the umpire, declares.

Soumit looks at the bleachers, Suman did not return the gaze, just showed the V sign.

"What an Ace!" Suman heard Sally cheering, Sally camp bursts into roar, "Go Soumit, go Soumit." Next three shots were all fantastic Aces. Soumit wins the first game.

Now it is Caesar serving, long and strong but right about the waist high and on to Soumit's forehand, and he returns with a fast cross-court down the line that clipped the baseline. Caesar didn't even attempt to return. "Love – 15." Caesar begins to show some emotions, he goes back to the serving line tries to hit an Ace, the ball goes straight on to the net. "Second serve." Oh, no, again straight on to the net. Love - 30. Next two serves were perfect Aces and Soumit had no chance of returning them. Score is 30 all. Soumit is little perturbed, looking down and pressing his strings.

Caesar goes to the serving line, oh, my! What an Ace! It touched the center line and swerved to the left of Soumit with a whizzing speed and the score now is 40-30. Soumit loses the game by returning the next serve, which was a top-spin straight on to the net. Score is 5-1, and Soumit is going to serve for the set.

Three top-spin serves, and a fourth Ace clinched the set. Score stands at 1-1. We are into the third hour of the match.

Chapter 5
The 3rd Set and Cramp Begins

Normally, there will be a 5-minute break when the game is tied, and players change courts. Umpire Chavez announced a three-minute bathroom break. Suman wanted to come down from the bleachers and talk to his son and offer a few words of advice and encouragements. Sally looked at him and waved a sign—stop. Suman obliged and didn't step down.

Soumit did not go to the restroom, he was just sitting on the chair that was placed next to the umpire's stand. He was wiping the sweat that has been accumulating on his forehead. He checked his strings, was satisfied, and did not bother to open his bag and pull out the second one that has a brand-new strings, just put on by Ned Bigsby, the junior coach at Sally's.

The final set is about to begin. Soumit will do the honor, he will serve Caesar from the east side.

"Love all, serve," the umpire announces.

"Oh, no." Straight on to the net. Suman covers his eyes.

"Second serve!"

This time it was a perfect Ace, the ball clipped the left far end of the baseline and Caesar had no chance of returning it.

Suman began to feel little relieved.

Two more Aces and the score is 40 - love.

"Darn it!" Suman hears himself. Straight back to back serves on to the net and now Caesar begins to show some emotions.

"40-15."

Another perfect Ace. Soumit wins the first game of the final set.

Now it is Caesar's turn. His followers started stomping the bleachers and chanting, "Caesar, Caesar, ho, ho, Caesar!"

"Quiet, please," a strong and bold voice from the umpire quieted the crowd.

"Fault," a foot-fault. Second serve, a rather weak top-spin, Soumit had no problem returning a cross-court doublehanded backhand that clipped the right side of Caesar's baseline and he had no means to return.

"Yes, yes!" and pumping of fist, a little expression of excitement, then he goes back to receive the next serve.

Caesar hits a perfect Ace. Soumit had no way of even touching it. Score is now 15 - all.

Oh, my! The game is over. Caesar hits three successive serve and to return the last one Soumit stretched his long legs to the far right of the baseline and barely resists falling.

Soumit, serving Caesar, one all.

Third game begins.

15 – Love, it was a slow top-spin and took Caesar by surprise, he was expecting an Ace to the left, but the ball spun away to the right. Very clever serve, Suman thought to herself. Sally camp, started clapping and Sally heard saying, "Attaboy," Soumit raises his racket and acknowledges the cheers.

Back-to-back Aces. Score is now 40-love. Caesar gets ready to receive the final serve for the game. Soumit reaches at the top of the baseline to send his winning serve of the game. Suddenly he starts to press his right hamstrings. He starts some sit-ups and gets ready to send the inevitable. And that was one, a perfect Ace to the left of Caesar, who had no chance of touching it with his extended racket. The score is now 2-1, Soumit.

Caesar changed his racket, Suman was waving to Soumit, asking him to do the same. Soumit looked at his dad, but didn't pay any attention, continued with the same one.

Caesar sends a hard Ace clipping the center line and as it was swerving to the right of Soumit, he tried to return with a stretched leg, but could not touch it. He comes back to the baseline to receive the next shot, but was limping and grabbing his quad. Apparently he is having cramp, that is the news buzzing in Suman's head. He has been wondering if Soumit had the banana before the third set.

"Oh, well, he is a fighter, he will definitely pull it through."

Caesar serves three more aces and tied the game with 2-2.

Now it is Soumit's serve. Sally comes next to Suman, asks, "Did he eat a banana?"

Suman just looks at her and shrugs, the answer is he doesn't know, because Soumit did not come out of the court when the umpire gave a three-minute break. So he doesn't know. Sally puts her palm on Suman's shoulder, trying to calm him down with some sort of assurance.

"Love all, two-all, serve." Suman could barely hear this from Mr. Chavez. His mind was fixated on his son's left quad. He must be having severe cramp, why did he try to stretch that hard to return that service? It was stupid. These words of his own were buzzing inside his head.

"40-Love. Game, Caesar 3-2," he heard the umpire.

Game 4 was spectacular, Soumit hit four consecutive top-spins, he had his racket little crookedly held (a slight variation of the grip that Ned taught him) and the ball takes huge top-spins and Caesar was nowhere to be found near it to be able to return. The last serve was a good Ace, but Caesar was able to return it, an easy volley for Soumit and the game was an easy love game. Score is 3-3, third set.

Cramp has begun to take its toll, Soumit has been grabbing his quads, and now his left thigh, he has been constantly limping and trying to ease the pain, it was very obvious. At one point Suman was raising his hands and asking Soumit to go to the sideline and tape his thigh, but Soumit did not pay much attention. Caesar looked at Soumit and just like a prowling cheetah would go back on the high grounds and position himself to jump on his prey. Caesar went back to the baseline with that sort of an attitude.

"Fifteen love."

"Caesar, Caesar, ho! ho!" Caesar, the Juarez boys bursting into roars. Caesar serves, tries to Ace Soumit, the ball was just about to chip away. Wait, Soumit, takes an impossible shot, two-handed backhand, and down the line to the opposite side and Caesar had no chance of returning.

"Oh, yeah, fist pumping." Score is 15 all.

"Go! Soumit, go!" Sally's bench starts cheering. Soumit raised his racket in recognition. Caesar regroups, goes back to his serving line, tosses the ball high and sends an awesome Ace, it clips the centerline and swerves past Soumit's stretched racket like a bullet. "30-15, Caesar." Next two serves were also Aces and Caesar is ahead, 4-3. Soumit now serving for the most crucial game of the match.

"15 – love," a spectacular Ace, Caesar has no chance of returning. Going for the repeat. Sally comes next to Suman and tries to calm him down.

"30 - love," a perfect repeat of the same. Sally bench goes absolute vocal. Soumit is composed and collected. This is what Suman has been telling his son all this time, when the outer world is full of excitement and exuberance, the inner world must be at peace like the middle of the ocean—unflinching and maintaining its constant motion. Very difficult, especially for a fifteen-year-old. But Soumit is showing such class to this crowd and being an example to his peers.

"30 - love." No sign of any emotion, just grabbing his quad and switching back and forth between the quad and the thigh. He is about to serve for the game now, and that will give him a 4-all position.

"Oh, no, not now," Suman was heard saying these words little louder, Sally puts her palm on Suman's shoulder. The ball hits straight on to the net.

"Second serve," again on to the net. "30-15."

"Soumit! Soumit! Soumit!" Sally bench tries to cheer him up.

A perfect Ace this time, and a repeat. Score is 40-15. He does some sit-ups and grabs his thigh again. Goes to the baseline. A big top-spin that totally catches Caesar by surprise and finds him nowhere near the ball.

"Foot Fault."

"40-15," "Second serve," the umpire announces a little louder, Soumit is about to serve for the game. He is limping, grabbing his quad and going back to the baseline to serve.

"Double fault, 40-30."

"Second serve, 40 - All, deuce." The next two serves went straight on to the nets and Soumit loses the almost-won game.

Score is 5-3, Caesar. Suman slumps and hides his face in his palms. Sally tries to comfort him by placing her left hand on his shoulder, but of no use. He cannot even raise his head and open his eyes to see the inevitable—his son losing the match, so deserving to win it. He is cursing inside, DAMM YOU, CRAMP, NOT NOW, PLEASE."

Caesar goes to the baseline to crush injured Soumit.

"Love - All, Serve."

"15 - love." Sally bench is absolutely quiet as if all are dead there.

"Fault, second-serve," that was a foot fault.

"Fault, 15 – all." Crowd is still in absolute silence.

Caesar tosses the ball high and sends a fantastic Ace, Soumit again stretched his right leg to the max to try to reach the ball with his forehand, the ball touches the top of his racket and bounces off to the right.

"30-15."

Another Ace and Caesar is now serving for the match.

"40-15, match point, Caesar."

And it is a perfect Ace.

The game is over. Soumit covers his face for a few seconds, approaches the net, shakes hands with Caesar in a very dignified way and leaves the court. Crowd gave both a standing ovation. The match lasted for 2 hours and 51 minutes.

Sally and all other teammates from the Sally camp come up to Soumit, Sally gives him a very big hug, other teammates give him hugs. Finally he comes to Suman, and they embrace each other, and none would let the other one go.

Chapter 6

The Ride Back and Cramp Taking Its Grip

The cramp has already set in. It is spreading from the upper-left quad down to the calf and even at his left foot. Poor Soumit is pressing his left quad and rubbing them hard.

"Dad, can you drive a little faster? It is getting unbearable. I am sure as soon as I reach home, Ma will make me a good bowl of chicken noodle soup, and it will go away."

"I am trying, Baba. Just try to take your mind away and concentrate on some other things. Say, we play a game of memory game. But this time I say two names and you reply with three, and the process continues."

"Okay," was a short reply from Soumit.

Suman pronounces two names: "Marco Polo and Vasco De Gama, your turn."

Soumit retorts, "Marco Polo, Vasco De Gama, Jane Austin, Shakespeare, and Lord Byron."

Suman repeats all five and then names "Rabindranath, Michael Madhusudan, and Vidyasagar."

Suman has this innate winning strategy, even with his own sons. He tried to play this trick on Soumit, for he knew that Soumit never heard of Michael Madhusudan; perhaps he heard about Vidyasagar and definitely about Rabindranath from his mom. But Soumit is something else. He has been blessed with a photographic memory and showed enough evidence to his parents from the age of two. He easily repeats all nine and then took his chances on his favorites: Pete Sampras, Andy Roberts, and Ken Roswell. Suman passes again. Now it is his turn for the next three. He sidetracked a bit, goes on to Hindustani Classical: "Bare Golam Ali, V. G. Yoga, and Ali Akbar." He thought he got his son now. No way. Effortlessly Soumit repeats all fifteen and then chooses science: "Pascal, Gauss, and Newton."

The game is getting tougher; however, the purpose is partly being served as far as Suman in concerned. He got his son's mind diverted from the cramp and on to a different level and this gives him pleasure. He tried to repeat all eighteen, thus far mentioned, but fumbled twice in the middle when Soumit invoked his favorites from literature. He surrendered. "Oh, yes," little pumping of fist. "How are you feeling, Babu?"

Bengali fathers, at least this one calls his sons Baba or Babu very affectionately at times when he needs to comfort their pains of any kind.

"I think I am getting better; it still hurts at the calf. At least it is the left calf and not the right one. Can we stop by at the gas station and get some Gatorade? I am very thirsty."

"Oh, yes, we will also get some ice, it will help your muscles a little, I will take the next exit."

"Thanks, Baba."

"How did I lose the third set? I was so determined; how could I make those two double faults in a row?"

"You want me to tell you how?" asks Suman.

"Oh, yes, please do."

This is the best part of Soumit. He always analyzes his failures. He always wants to know why and why he cannot reach at the top.

Suman starts: "Remember your lessons, Sally, Ned Bigsby all told you, and I repeatedly pointed this to you, toss the ball high and straight, then you will have the full view of the ball and head of the racket to place the ball wherever you want. I don't know why, on both occasions you tossed it way to your left and you had to bend over to your left to serve the ball. This motion is not elliptical, and it does not give you all your strength to hit the ball at your hardest. Perhaps you were already having pain from the cramps. Remember, son, as I told you in Mahabharata and Geeta, Lord Krishna taught Arjun the perfect way of execution. Even though Arjun's teacher Dronacharya sided with the Kauravs, it was Lord Krischna asked Arjun to stay absolute focused with his task. Any vacillation from that would not have earned Arjun the reward of win, however destructive that might have been. When you are at a task, it is not the end result that you look for; your charge is to perfect the process of execution, for you have no control over the outcome. We call it – 'Ma Falesu Kadachane'- that is never to focus on the result, but always on the process. You may aim for the desired result, but your task is to perfect the process of executing it."

"Baba, take the next exit, it is just coming and there is a Shell gas station."

It is the amazing photogenic memory that baffles Suman often times about his son. Once they were talking about changing car tires and Suman was telling his wife that he needs to look for some rebates or discounts. Soumit was quietly doing something mischievous in the back seat with his little brother. He just said, "Baba, there is a discount tire store in El Paso, just right of the Mesa exit."

"Wow! How do you know?" Suman's eyes were glittering.

"Oh, I just saw it the other day when we were going to Dadu's house."

Dudu was Mesomashai, Dr. Bhaduri, who is like parents to Suman and his wife and the boys call them Dadu. Soumit was only seven years old back then.

Suman stops his pick-up to the far right of the gas station. He has enough gas to go home. He just needs to pick up a bag of ice and some Gatorade to get the fluid flowing through the cramping leg and put lots of ice on his son's right hamstring and quad. He told Soumit not to move, keep his leg up straight, almost mounted on the dashboard. He managed to buy a small bag of Cheetos, a bag of ice, and two Gatorades. Then it donned on him, how is he going to put the ice on his thigh? Back of his leg? He went back inside the store, came back with a small bag of plastic bags. He hurriedly went back of his truck, slammed the ice bag on the bed of the truck, and got bigger chunks broken into small pieces, enough to have three little plastic bags filled. He put one bag on Soumit's thigh, and asked him to put the other bag under his hamstring. Opened the lid of the Gatorade bottle.

"Are we ready, son? Let's get back."

Started his engine and veered to the left and got on the feeder road to get on the I-10 West. It will take them another 30 minutes to go home.

"Hey, drink some Gatorade and press the ice bag gently on the thigh."

"Okay, Baba. Thanks."

This is typical of Soumit. He never forgets to thank people for any little thing that they do for him.

The pick-up was cruising at 76 miles an hour, slightly above the speed limit on I-10. Suman looks at his son, it seems the ice bags are working, and he is little relaxed. Suddenly, Suman hears Soumit talking to himself. He is in a trance-like situation and half asleep.

"Why did you have to go for that volley? You idiot! You can never perfect that shot, you think you can play like Pete? Hah, stupid, idiot."

Suman was listening to the soliloquy so deeply that he was about to veer off the road, he jolted the pick-up back to its lane, fortunately no other car was passing by. This sudden jerk of the vehicle awoke Soumit.

"What happened, Baba? Why did you change lanes so abruptly? Are you sleepy?" All the right questions with full of anxiety and caring tone.

"No, no, I am alright, just a little dozing off, it will not happen again. You go back to sleep. No, I am not sleeping, just thinking about what could I have done to save the third set."

"Don't worry about it. You gave your best shot. Move on. Try again next year. You still have one more year at the school. What time is your lesson tonight?" Suman tries to divert his attention.

Soumit plays violin. He is the concert master in his school orchestra, he also plays with NMSU Symphony. Dr. Gabbi, the conductor at NMSU, just enrolled three young players into the university symphony since last year. Soumit was so happy to be called by Dr. Gabbi herself after two annoying nights after the audition. Their practice sessions are on every Sunday from 3-5:00 P.M. They are preparing for a fall concert, a two-day gala event at NMSU Music Recital Hall, for fundraising and meet-and-greet. An annual event, highly publicized and meant to raise scholarship monies for budding music students. Last year they raised 50,000 dollars. Soumit did not play in that for he and two other high schoolers got inducted only this summer. They will be performing this fall with a renowned guest artist by the name of Nadja Solemberg, an old student of Dr. Marianna Gabbi when she was in Philadelphia and playing for their symphony.

"Oh, our school rehearsal is at 8:00 P.M., am not worried about that at all, but I am little bit worried about tomorrow's practice with Dr. Gabbi, you know how critical she is!"

"Yes, I know. Just practice a little bit in front of your mother before you go to Dr. Gabbi's practice. Your mother has a perfect ear and she is equally critical of any wrong notes and bad intonations."

"I know, she is the toughest. I always freak out practicing in front of her."

"But you know she wants the best for you."

"I know that, Baba."

La Mesa exit was coming fast, Suman must take the next exit to go home. As he was pulling the pick-up in his driveway, Soumit saw his brother and two other neighborhood kids were placing the goal posts for their rollerblade hockey.

"Hey, guys, just wait, I will be joining you."

As they pulled up the truck inside the garage, Soumit jumped out of the passenger seat and called his mother, "Hi, Mom, I am back."

In a few minutes he was ready with his gears and on the street with his hockey stick asking for the puck from his brother to take a free hit.

Suman did not close the garage door. Just looked at his sons. Heaved a deep sigh of relief and went inside with a little smirk on his face.

The Bicycle Kick

Chapter 1

The Phone Call

"Chhutka, here is a call for you, I don't know who that is, come down and pick it up."

Suman, Chhutka's (nickname of Krisch) father, was calling for him. Krisch or Chhutka just got back from school, a new one for him in this unknown and yet unfriendly town, threw his backpack on the sofa and went straight to the upstairs loft at their rented house. He was deeply immersed at the new Gateway desktop that his parents just bought for his older brother, who is just about to be shipped out of the house to his about-to-begin college life in Chicago.

"Okay, Baba, I am coming, ask him to hold on, please."

"Hello, who is calling, this is Krisch."

"Hey, this is Matt, assistant coach of Brentwood Eagles. You made the team, Coach Wisermann wants you to come tomorrow at 4:30 P.M. for practice, we don't know what position you will be playing yet, that will be determined by the coach at the end of our first week of practices, can you make it?"

Krisch nonchalantly just said, "Yes." Actually he was very happy, he was missing his Owensboro Nightriders, where he played for two years and was one of the top scorers. He played center-forward position and occasionally right half-back. This "moving" has been deeply bothering his eleven-year-old body and mind. But he never said anything to his parents and not even a word to his older brother, who is his biggest supporter and mentor.

Baba was impatient, both him and the mother were anxiously waiting for this decision, they knew how much little Krisch is missing his soccer.

"What happened? I know they selected you in the team, what position? When do you start? So many questions from an impatient adult!"

"Practice tomorrow at 4:30." Just a very short answer.

Chhutka runs upstairs.

The Gateway desktop was not properly configured yet. Dada, Soumit starts his first year at North Western next week. His orientation is on Friday and today is Thursday, the week before. He hasn't come back from his rehearsals from Blair Music School at Vanderbilt. Even though Blair offered him a full scholarship for the entire four years of his undergraduate studies there, but he was set to go at NU (North Western).

"Hey, bro, were you able to configure the system the way I want?" First thing Soumit asked as he came back from the practice and went straight upstairs.

"Not yet, working on it. It will be done before dinner."

Krisch is running some "fixes" that Microsoft provided with its Windows operating systems. This operating system is new to Suman, he was used to the old DOS-based system; however, to Krisch's eleven-year-old mind all new things are worth taken apart and made them to work as they should. Yesterday, they (the brothers) installed additional RAM, now he is running some patches to make it run faster and more efficiently. Suman used to freak out at this unprecedented reach of Krisch towards all things electronic. He used to scream and call for his wife. Neela, Suman's wife, used to get perturbed, and scold Krisch profusely, but it was Soumit, or Shoumit, as he would like to pronounce his name with an accent on the "h," who told his mother not to.

"Ma, Bhai (younger brother) is a natural, don't stop him. He will be someone, someday in this field." That advice calmed her down.

"Hey, guys, come down, dinner is served," Baba called for the boys. He was helping his wife to set the table up. Neela is a great cook, she always makes something special for her boys, tonight it was mashed potatoes, baked chicken, and rice pilaf. Of course salad has to be there. Since it's a weekday, no wine for the adults. She was very happy; the family is back together after almost a two-year break. Boys and their mother in Owensboro, Kentucky, and Dad back in Las Cruces, New Mexico. With a new job and juggling two boys singlehandedly she was exasperated. When this new opportunity smiled on them, for both of them to be professors at the same university, they could not say no. Obviously Soumit has a lot to do about it. He took care of all the details about his brother's wellbeing—school, soccer club, cello practice at youth orchestra, etc.

Brentwood Middle closes at 3:20 P.M. and by the time Krisch can come back home and get ready to go to his first day of practice it may be too late. Suman got back from his office early around 3:00 P.M. and got his star soccer player's bag ready. He was thinking ahead, got some chicken nuggets from Chick-fil-A and his water bottle with lots of ice. He was at the pick-up circle for parents at the school around 3:25.

"Baba, let's go home quick, I have to be in the field by 4:25."

"No, we don't. I got everything ready for you and your food. Let's go to the field. There are good bathrooms there at Crocket Park. You can change there, I got your soccer cleats, and last year's jersey from Owensboro. They will give you new jerseys, I believe. So wear these for a few more days."

Chhutka didn't say a word. He was actually very happy to see that his father brought everything for him, and he is ready to just play.

Chapter 2

First Day of Practice

Krisch has been wearing jersey number 10. He has been wearing this number since the age of six, when he used to play for Little Dragons in Las Cruces. He never thought that he may not get that number. Number 10 in soccer is a special one. All greats wore number 10, Pele, Maradona, Ronaldinho are the most notable ones. He loves Ronaldinho. He and Baba always argue over who is the better one. Baba, of course, goes for Pele, the legend; however, Chhutka will not change from his favorite Ronaldinho. Nonetheless, he is not ready to give up number 10. Let's see which number he gets this time. He wore his blue and white number 10 from Knigh-triders for his first day of practice. All fifteen players were there in the field. Brentwood Eagles is a division-1 team under the age of thirteen in Tennessee. Last year they won the State Runner-Up position, Coach Wiser-mann is highly respected in the circuit and all parents respect him as well. He is stern but gentle with his boys.

This year Brentwood Eagles had to recruit five youngsters because five of their star players moved up to high school and can't play in this age group.

Both Tim and Matt are already in the field. Matt is lining up the boys for some drills. Suman and few other parents found some benches to sit and watch. Crockett Park has eight fields, full-size soccer fields. As a matter of fact the city is clearing up some more trees and preparing the grounds farther to the west of the fields for six additional fields. They have plans to put bleachers at all fields. Next year they are going to hold statewide championship tournaments along with two fields in Memphis and Knoxville. Soccer is getting very popular all over Southeast. Suman heard from a parent that there are over 100 teams competing for the Tennessee Championship Trophy this year.

"Oh! They have started running, this is what Tim does every year. He makes them run for 20 minutes first, about two miles, and then lets them play one-on-one for another 20 minutes and then groups them in two teams and lets them do a scrimmage. Hi, I am Leslie, I am Brian's mother, you must be Krisch's dad, right?"

"Yes, I am, nice to meet you, Leslie."

"You know Brian and Krisch are in the same section," Brian's mother says."

"Oh, Mom, Krisch is so smart, he can fix any computer. They are becoming good friends."

"I am glad that they are—we just moved here from Owensboro, Kentucky, and Krisch is still trying to adjust, we all are," Suman says politely. "What position Brian plays?"

"Oh, don't worry, Brian plays right half back and no one is going to take his old position, I understand Krisch plays center forward?"

"Yes, he has been playing center forward mostly, sometimes right wing. But he prefers center forward."

"That's great, we are in need of a good center forward. Chris just graduated from middle and moved on to high school. His position is open, and it is perfect for Krisch."

"Hi, Jess, when did you get here from New York?" Leslie approaches Jess, another soccer mom.

"Look, there is Brianna, Corbin's mother."

Ladies got to their circle. Suman was watching all the boys and their readiness to play soccer. Few of them were already too tired since they are coming right after school, perhaps hungry. Coach Wisermann is not running with the boys, his assistant was running, though. Tim was looking through a chart, some kind of a roster. Suman was sure that Coach was deliberating in his mind how best he must fill the empty positions with the players he got.

"Listen up, after the next round, you all go to the far end of the right side, next to the restroom stalls." He just shouted at the slowly running team.

"Yes, Coach!" many of them answered back.

Suman was in his usual restless state, he was pacing along the perimeter of the field. He was trying to measure the coach and the assistant coach and their assessment about his son.

"Okay, boys, let's split up and do a scrimmage for 20 minutes. Team A, Brian, you are the captain, go ahead and select your eight players, Corbin, you are the captain of Team B, select seven players; let's start the scrimmage in three minutes."

Suman watches. Brian selects Krisch. The goalposts are moved forward, these are moveable goalposts. Each team has three forwards, two half-backs, and one full-back and the defender. However, Brian's team has an extra forward and that is Krisch, playing his favorite position—center forward.

"Oh, my! Goal."

Krisch took the ball from Brian and ran so fast with it that no one can even come near to dribble or take the ball away from his feet. He dribbled past Corbin and gently pushes the ball into the small net.

"That is fantastic," Brian's mom exclaimed and gave a thumbs-up sign to Suman.

Corbin's mom did not say a word. She was not expecting Corbin to be beaten that easily. Corbin is a star goalie. Last year he saved thirteen goals that were almost inevitable and won a "Best Goalie – under 14" trophy. Brianna is very proud of her son, and this little defeat by an unknown quantity hurts her feelings. Scrimmage continues for another 15 minutes. Krisch scored two more goals, his team won 4-2. He was the highest scorer for that day.

"Hey, guys, listen up, all assemble over the far corner in the shade, next to the restrooms, we will be distributing the jerseys and go over some basic rules," Assistant Coach Matt almost shouts at the boys. This is how the assistant coaches always behave, as if they are going to be usurped by their bosses, so they have to yell.

"Disgusting," Suman just murmured and found himself in front of Laura, another soccer mom.

Laura just smiled. He felt little embarrassed and smiled back. Krisch got his favorite No. 10.

"Dad, they did not tell me what position I will be playing, why? Did I not play well? I scored three out of four goals for my team."

"I don't know, son, I hope they will give you, your preferred position. Let's see."

Dinner was simple, pure Bengali dishes, Rice, Daal, some fried vegetables and chicken curry. Krisch likes Mom's chicken curry; however, his brother always complains that it is too spicy and hot.

"Hey, bro, so did they give you your center-forward position?"

"Don't ask."

"He is not happy about it."

"No, not that, they are still deciding. I will score more goals in the next scrimmage."

"That's my boy. Score five goals. They will have no choice but to give you your position, you so deserve it."

"Oh, yeah," big bro gives him a thumbs-up.

Next practice is on Sunday at 3:00 P.M. It is a two-hour practice, sort of a warmer exercise before their first friendly scrimmage against Franklin Fire, also a champion team. Brentwood and Franklin have been archrivals for years for no apparent reason whatsoever. When Krisch was signing up for the tryouts to join Brentwood

Middle, Krisch heard that every soccer game or a football game between these two will see a commotion, even fist-fight among players and even parents. Brian told him that all the boys from Franklin Fire hate the Brentwood boys. One reason may be most of the Franklin Middle or Franklin High kids come from relatively poorer families and Hispanic or African-American origin. Even at that early age they have learned the distinction of class and race and innate discriminatory hate from parents and surroundings creep into their young minds. Krisch does not belong to either group, so he should be safe. But not!

"Line up, guys," Matt, the assistant coach, shouts out. "Today we will play one-on-one first. We will split the field into four quadrants and each of you play against your opponent for the seven-minute half. We will then have our two teams selected and have a full scrimmage for 30 minutes; before that just run four laps and take your position in any quadrant."

Four laps around the field is about a mile, that should give enough warm-up before practice. This is nothing new to Krisch, he is used to a two-mile run plus all the scrimmages during a regular practice. Only time he has some trouble when he gets some allergy. He has a dry allergy, which he has inherited at birth from his mother. He was born in New Mexico's semi-arid dry weather. But moving east helped him quite a bit. He does not suffer severely from asthmatic conditions anymore, but occasionally gets whizzing, so his mom always asks him to keep the inhaler with him. She made sure that it is in his soccer bag just in case he needs it.

"Okay, boys, Coach will be little delayed today, but he will be here before the final scrimmage. We will have our positions selected after the scrimmage and the roaster for our league when he gets back from the league organizer's meeting. Corbin, you take the first quadrant, Krisch will play against you. Remember, guys, you are playing for all positions including goalie. You play two seven-minute halves. We have four teams starting now and the other four teams will come with me and do some more exercises. Let's begin." Matt says all these in a single breath.

Krisch is ready. He must score at least two goals against the best defender in this division. These seven-minute halves are played without any referee, it is some sort of self-monitored, Matt is the roving referee, he runs in between the quadrants and monitors to some extent. In quadrant 1, Krisch versus Corbin, next to it is the 3rd, where Brian and Doug are facing each other. The other two quadrants are taken by Chris and Bill and the last one is being battled by Devin and Josh. Rest of the boys went to the far side and are doing some more workouts as suggested by Matt.

Corbin pushes the ball from the center and tries to dribble Krisch, he gets past and kicks it to the goalpost. No, it did not come near the little goalpost. Krisch runs to the ball and controls it with his right foot, runs to the left of Corbin, as if he is going to pass, Corbin charges back but could not even touch the ball or Krisch, Krisch runs, boy, he can run! Oh, no, Krisch gently pushes the ball through the goalpost.

"One – zero, Krisch," Matt announces.

Neither Krisch nor Corbin noticed that Matt was not refereeing but watching closely, with his notepad on the clipboard. He scribbled something; it must be the score.

"Carry on, Krisch."

Matt moves closer to field/quadrant number 4. Corbin picks up the ball and kicks it forward but could not get to it in time, Krisch already has full control of it. Dribbling is not the best of Krisch's strengths. He is not selfish, he takes the ball, runs very fast and centers it to the far wing or gives a short pass to the following half backs, either to his left or right, and makes room for himself in front of the goal to score. This has been his typical roadmap. He learned this strategy back in New Mexico, where Coach Luis taught them offensive moves since he was six years old. When he moved to Kentucky, Coach Mo asked him to change, and he tried but failed miserably and gently went back to his old style. He was the highest scorer in their division with 24 goals in last year's tournament and owned the Owensboro Open League. Today is no different. But there is no one to pass to, so he makes up one imaginary, following him as Corbin comes charging him. Corbin is tall, almost six feet one at that age, Krisch is only five feet seven, he is only eleven, yet to grow up to his potential. Corbin cannot run very fast, Krisch goes a little back and picks up the ball and again to Corbin's utter dismay takes a shot at the goal from 20 yards. Clear goal.

"Two-zero Krisch," Matt again declares.

First half is over. Time for the second seven-minute half. No break, just change the sides. This half was even more eventful. Krisch scored two more goals and conceded only one. So he wins 4-1. Matt tells these eight boys to rest for five minutes and then go back to the far side near the restrooms and wait for him, he needs to get the other four teams started. But there is one player short, so he will play against Jim. Jim Rodriguez is a stocky boy, looks very tough and always gets into some kind of fight with opponents, however, scores one or two goals in every match. Matt was playing against his team. Jim scored three goals in the first half and scored another one in the next half. Scrimmage ended.

Time for Coach Wisermann to address the boys with full details of their roster of practices, matches and the positions each player will be playing this season. From next practice on, everyone will have a full 30- to 45-minute scrimmage besides endurance training, team building and leadership and motivation development. They will work off the field for a couple of Sundays to build homes for Habitat for Humanity.

“Okay, boys, drink some water, go to the bathroom, or whatever, we will meet at the far side in five minutes. It will take 45 minutes and we will then distribute the roster and the tournament schedule.”

Ride back home was non-glamorous. Krisch got his well-deserved center forward position and the first match in the tournament is in two weeks’ time.

“Hey, bro, did you get your position?”

“Oh, yes!” Krisch runs upstairs, where he has some

unfinished business with the new Gateway desktop. It is not quite ready yet and dada (older brother) leaves for his college in five days, on Friday after Labor Day. He must go on the 12th or 13th since his first day of class is scheduled to start on the 15th of September.

"Chhutka, come downstairs and make the table ready for dinner," Ma calls for him.

"Let me help you, he is trying to get the computer ready for dada," Suman intercepts. "They will come down when we are ready."

Not much of chitchat at the dinner table, everyone is consumed with the big brother's move to Chicago that is just coming around the corner.

"Hey, guys, I leave for D.C. for a meeting tomorrow and will be back on Wednesday. We will be leaving for Chicago on Friday, that way, Soumit will have a couple of days to get acclimatized there before school starts." Such was the plan at that moment.

"Why are you going there again, Baba?" Krisch asked.

"I think I told you already, I have a presentation on the US-Mexico border environment, particularly air pollution before USEPA, World Bank people. EPA funded our project, remember!"

"Oh, that one!" Krisch said very nonchalantly. "Yes, Baba, did you not collect data at the San Diego – Tijuana border at the border crossings and survey drivers of the cars that were about to come to the US?" Soumit asked.

He was very proud of his father doing these kinds of research at some of the exotic places. Suman remembers when he was giving a talk in Ahmedabad at the Nirma Institute of Management on Environmental Economics, he and his mother were in the audience, and his big pair of piercing eyes were following every movement of Suman and Suman could feel those looks in his bones. He was extra careful to make his accent more Americanized.

"Boys, please keep the plates in the sink, put some water on them, Baba will rinse them and stow them in the dishwasher."

"Okay, Ma, I will do it," Krisch volunteered.

Neela and Suman exchanged glances.

Next two days soccer practices were uneventful. Neela had to juggle between her teaching schedule and dropping and picking up Krisch from his school, his cello rehearsals, and yesterday's soccer practice was a bummer. Krisch thought it was at 4:00 P.M., and overshoot it by a half-hour. He was delayed by half an hour and could not start. He was very upset, however, Coach Wisermann talked to him at the end of practice and calmed him down. Dad is supposed to be back tonight (9/11) and everything should be fine. But someone had a different plan.

Ever since 9/11 happened everything changed in America. Suman got back home two days later, Soumit barely made his first day of class at NU. No soccer practice for a week, schools were closed for three days, and ten days later life just began to normalize. At home there had to be many adjustments, big brother is now off to college, the entire family den is all available to Krisch, and got messier.

"Baba, we need to go now, hurry up."

Suman did not go to office today; he was too tired from all running around. He just visited the construction site of their new home being built and got somewhat unhappy because there were lots of little mistakes. So that will cause some more delay for moving in. They were supposed to move in during the Labor Day weekend before Soumit left for college, now it looks like another seven to ten days will be required. Neela was furious, she has been spending more time at the office lately.

"I am ready, and the car is outside, just get your soccer bag and the water bottle, let's go."

Krisch jumps into the car and sat next to Dad. He is not a talkative boy unlike his brother. He is much introvert. Suman kept quiet. They reached the fields.

"Go have a great practice. I will be back in an hour."

"No, stay here, please." And he jumps out of the car and ran to the field, where some boys were already trying to do all kinds of tricks with their soccer balls.

Krisch has his own, but he did not pull it out of his bag, rather he took a couple of sips of water from his bottle and starts warming up. Today's practice would be very demanding since they missed a few practices. League matches will start from next week and they may have two games per week. The expedited schedule has been given to all the coaches and everyone agreed that this roster can be maintained without much of a problem and the tournament can finish before the spring semester begins.

Krisch scored three goals in the first scrimmage but could not score any for the second one. Each was 30 minutes long. He attempted a few special things this afternoon—a free kick that was supposed to take an outswing and take the upper-right corner of the inside of the crossbar and go in, that didn't happen; it did take the outswing but it was little too much and missed the goal by a couple of feet. No goal there. Then at the 19th minute of the second scrimmage their team got a corner kick since Corbin had to punch out the incoming ball and it went over the goal line. Brian took the corner kick aiming Krisch's head, Krisch was right there positioning himself juggling between the opponents, but as the ball came a little higher than he anticipated and attempted his bicycle kick and missed it. It was an awesome try.

"Wow, wow, what an attempt," Brian's mother screamed. "Attaboy."

Ride back was grim. Suman tried to make things little lighter. He knew that his son is very disappointed—first he missed the free kick and then he missed his most favorite bicycle kick. "Hey, Baba (Suman calls both his sons Baba intermittently to comfort and console them), don't worry, you will definitely do it next time."

"No, Baba, we do not have practice tomorrow, but I want you to come with me in that little field next to Hillsboro Pike and make me practice and perfect this stupid bicycle kick. I must be able to do it every time there is an opportunity, can you do this for me, please?"

"Yes, we will do it, but it will be after I come back from the office."

"Okay."

Next two days Krisch practiced with his dad. Suman kicks the ball from the corner of the field, fortunately that time of the day, around 6 P.M. almost dark, all the little boys and girls have gone home after their soccer practices, there is still about fifteen to twenty minutes of fade sunlight. This has been the ideal time for the son and father practice. On the first day of their special practice, Krisch scored three times out of ten tries. Of course, Suman could not kick the ball in the desired spot and the angles were not there either.

Suman was never a soccer player. He played some in his childhood just for a couple of years and then as he got his eyeglasses, he quit soccer and picked up the cricket bat, and he was quite good at it. However, he has been an avid fan of soccer, and has a pretty good sense about all the aspects of the game.

"Hey, let's pack up for today, it is getting dark and almost 7 o'clock. Don't you have homework for tomorrow?"

"Oh, yeah! And I have cello lesson tomorrow and I must practice tonight."

First match is next week Saturday at 10:30 A.M., then they will play again at 3:30. They will have to play one match also on Sunday, all at the Crockett Park soccer fields. Next week, however, they have to go to Hendersonville, about 40 miles from Franklin. So the coaches scheduled three practices this week, Wednesday, Thursday and Friday.

Next three days passed in a hurry. Today's practice will be challenging. They will be having two scrimmages, each 40 minutes long with only a five-minute break. Krisch is ready to try his bicycle kick again, at least that's what he told his big brother on the phone last night. Dad overheard the whole conversation. This is really amazing to both the parents that the brothers even though they are almost six years apart can talk on any subject for hours. Big brother was saying, "Bro, try and try regardless you fail, I am sure you will make one fantastic bicycle kick in the match, but practice and practice, remember what Ma tells us." Big brother and his advices have profound impacts on Krisch. He regards him more than his parents virtually on everything.

Scrimmage just started, Krisch is playing opposite Brian, Tommy the great full back and Corbin the number-one goalie for last year. Coach Wisermann personally selected and put the boys in halves in such a way that today's scrimmage will test their ultimate skill and endurance. Krisch seem to be not bothered by this selection, he is just all about his business. This apparent maturity comes from his working solely on the computers. He knows what needs to be fixed and gets on with it without much reverence to the adversities. Dad and Mom both admire this trait of his.

"Hi, Suman, how are you? Did you notice they put Krisch in the weakest side, I guess just to test him out for the last time before this weekend's match, don't worry, he will be fine." All these words in one breath. The speaker was Brian's mother.

"I know, and I think Krisch will be fine, thank you for your trust, though." Just a short reconciliatory note from Suman while his gaze was on some other place.

Krisch had some different plans for today. He was determined to score goals and that too in a very special manner—the bicycle kicks. Match has already started, however, as if he was not there, he was not keeping the ball at all just passing on to the wings. As Scott was trying to dribble Brian and pass on to Krisch since he was just ahead of him, Krisch has already moved away from the position where Scott was about to pass, but he was already committed and did the obvious. Jake got the ball and passed on to Kyle and Kyle dribbled three players, including the defender, and scored the first goal. Everyone was looking at Krisch as if it was his fault they are down by 1-0. Next ten minutes were only up and down, no one scored, Krisch took a shot from 20 yards but Corbin was right there to punch it out and as he did it, he let a big yell, "Oh, yeah!" Krisch did not look back.

Halftime break was only for five minutes, he came out of the field and drank water from the bottle and Dad had a banana for him, this was big brother's advice. Banana gives instant potassium in the body and muscles get their much-needed juice. He had two bites and had another gulp of water and off he goes. As he was running back to the field, Brian waved at him, he replied and took his position at the center.

Suman cannot seem to understand why these guys kick the ball backward when they start. He thinks it puts them in a defensive mode rather than offensive, he told this to Krisch and Krisch's answer was "Dad, it is the American way." Well, go on with your ways, boys. Such was his thoughts. But he must watch his son scoring at least one goal. It is very important for him. He was watching Krisch just dribbled Brian with his famous one-two pass—attempt to dribble but don't complete it just pass it in front of you and rush on to the ball. He just did that, and Brian is now way behind to catch up with him, he moves to his left and dribbles another back, Corbin is getting ready to defend the citadel. Oh, no, Krisch attempts to get past Corbin by attempting the same technique, but Corbin was too tall for him and dived straight on to the ball and captured it. No goal. Poor disappointed Krisch. "Oh, that was a great effort." This time it was Laura, Corbin's mother. Suman just gave a gentle nod. Ten more minutes left in this first scrimmage. The score is 1-0, Corbin's team was winning.

Watch out, the ball is now in Patrick's left foot and he is advancing fast, Brian is rushing from behind to tackle Pat, Krisch was right there and Pat passes on to him, Krisch has a difficult shot but he can still take a shot at the goal. He made a quick move to his right and posed for a shot, Brian comes on his left and he took a ground shot to Brian's right side. The ball caught Brian's right stretched leg and went over the goal line.

Corner. Suman now understands his son's plan. It is a master plan, but will it work? He kept his thought to himself and tried to calm him down by keeping his fingers crossed. Patrick was ready to take the corner kick. Krisch approached Patrick and said something,

Suman only knew what was it about. It was the type of the corner kick that he is looking forward to score big with his bicycle kick. Patrick sets the ball for the kick. All eyes were on Krisch, Brian himself was guarding him. Krisch moved a little back and waves at Patrick. Pat takes the shot, not on to the top of the goalpost for someone to take a head but about five yards inside the box, almost closer to the penalty shot marker. Here the ball comes with a perfect trajectory, Krisch was right there, the ball is approaching from right to left, before it goes on to Krisch's left side and touches the ground, something grand just took place, Krisch is almost parallel to the ground, his one hand fully stretched was on the ground and with his right leg forward and left leg half crossed took a big shot. The ball travelled seven feet left of Corbin's stretched arms and into the net. It was a perfect BICYCLE KICK.

"Oh, my goodness, what was that?" Corbin's mother almost closed her eyes with both palms.

"That was fantastic. I have never seen something like that," Brian's mother complimented.

Suman closed his eyes and looked straight ahead, and their eyes met halfway. The scrimmage ended with 1-1, tied. At the end of the practice Coach Wisermann waved at Krisch. He waved back and arranged his soccer bag and deposited it to Suman.

"Baba, Coach wants to talk to me, I will be right back."

"Okay, I will be right here."

Suman waved at the other parents, slowly they were going back to their cars. He could see Coach Wisermann put his right palm on Krisch's shoulder and asks him to walk near the north side goalpost where Krisch took the winning bicycle kick. He was talking to Krisch, but Suman couldn't hear from that distance. He further asked Krisch to go near the spot from where he took the shot, and then pointing out to the corner. It was quite a discussion for about seven to eight minutes before they returned.

The ride back home was silent. Suman knew if he asks his son anything, the reply will be simple. "Nothing." So he just kept on driving until he pulled the car inside the garage. At home there was more silence, Neela is not there, she has been attending a conference in Denver and will not be back till Wednesday, and the big brother is in Chicago already started his classes. Suman was supposed to warm up the already cooked and neatly refrigerated food that Neela has prepared and left, he needs to only make the rice, that too in the rice cooker. He knew Krisch will take at least 20 minutes for his shower and by that time he can make the table ready for the two of them. Dinner was simple, rice, some daal, and cauliflower and potato lightly fried but no gravy and then of course, Krisch's favorite dish, almond chicken.

"Hey, Baba, come down, dinner is ready."

They ate almost quietly.

"When is Ma coming back?"

"Wednesday, around 7:00 P.M."

"Oh, she will then miss my first two matches this weekend."

"I guess she will."

"So what was Mr. Wisermann saying?" Suman knew the answer, still asked.

"Oh, nothing, just showing me the right place from where it would be easier for me to take the kick, he was not too happy that I purposely kicked the ball to Brian's leg and got a rebound and got the corner. He said not to do it because the other player might be more careful and in that case he will get control of the ball. I should have kicked the ball straight on to the left side of the goalpost and scored the goal or just passed on to the wing."

"So what did you say?"

"I said thank you and I will keep that in mind."

"Did he say anything about the goal?"

"He said it was 'spectacular'!"

"Do you have any homework for tomorrow?"

"Yes, just some writing. Couple of essays. I will finish them and then do some cello practice."

"Okay, son, I will be in the office room, come and get me if you need anything."

Next day's practice was uneventful, Krisch score two goals within first ten minutes and then played totally differently, he just kept passing to his teammates up and down, and to the right wing and left wing, and helped Patrick scored another goal. Today their Goalie is Corbin, and no one could get past him, so the score was 3-0 at the end.

"Tournament begins in two days, tomorrow is Friday, I want all of you to take rest and not touch the ball, just think and meditate a little bit how you are going to position yourself throughout the game. We are going to play against Chattanooga Rebels, and they are not bad, little bit aggressive, so you all need to play your positions firmly. So that's all for tonight, we will meet here at 9:45 A.M., our game starts at 10:30 A.M. Until then." Coach said all these without even pausing for a breath.

Everyone said "goodnight" and left for their respective cars.

Suman already ordered pizza, breadsticks and salad from Pizza Hut, just have to pick up on the way home.

Krisch woke up at 7:00 A.M. and already in the shower. Both the brothers must take shower first before they come out of their rooms. School days! Suman understands, but even on the weekends, same routine? This new house is rather big—four bedrooms on the second floor, and two more unfinished rooms in the basement and the boys have their own attached bathrooms and the master suite is quite spacious. Suman could not finish the basement before they moved in, that can be done slowly, he can hear the water running, it will take Krisch about 30 minutes to finish and get to the breakfast table. He quickly sprinkles hot water on his face and gets ready to cook breakfast for Krisch and himself.

Breakfast is simple, pancakes and bacon and some fruits and lots of orange juice. He checks the pantry, there is enough pancake mix and a brand-new maple syrup bottle; these things Neela keep handy and right in front of his eyes for he cannot find things even though they are right there. Neela teases Suman, saying, "It is just looking at you," and smiles.

Chapter 3

Tournament Begins

By the time breakfast was done and Krisch got his water bottle and extra pair of socks and a banana in his soccer bag it was about 8:30.

"Baba, we must go now, Coach asked us to go before 9:00 A.M."

"Oh, yes, I am ready, just go to the van, we are taking the van today, I forgot to fill up my car yesterday, and we will not have time to take gas on the way."

"Okay."

They left home exactly at 8:40 A.M.

"Hey, Krisch, come on over here." It was Brian. Krisch waved at his dad and proceeded to his left where Brian and a few other teammates were standing and having small talks and just waiting for the coach or his assistant to show up. They will play on field number 4, where a game was already in progress. They will finish at 9:30 and Kirsch's game will start at 9:45. These schedules run pretty much on time. Suman says hi to Laura and few other moms and dads. For a change there is Corbin's dad, Brian's dad, Chuck is also here today.

"Howdy, howdy." He just throws out the common greetings to everyone.

Laura started to respond, then saw Coach coming and totally changed her tone, "Look, guys, Tim is here and the boys are going toward the field, you should all go there and get our spots."

Everyone seems to agree and starts moving in the direction of field number 4.

Wisermann has a unique strategy—he does not announce the final 11 until the last minute. All fifteen boys are ready to take their positions and at the last minute, just before the referee whistle is about to be heard, he announces the defender's name and then the defense, and finally the offense. Today was slightly different. Looks like he already gave the final list on his clipboard to his assistant Matt. And that is Matt's dirty job to delist four boys from the starting lineup.

The first match was an easy win for the Brentwood boys, Krisch scored two goals and one assist that also resulted into a goal. Chattanooga guys could not penetrate Corbin's citadel. The score was 3-0. The next match is at 4:30 P.M. Suman wanted Krisch to come home with him and take some rest, but Krisch wanted to go to Brian's and play some new video game that Brian just got. So he bids a bye, saying, "I will be back around 4:00 P.M."

Time is now noon. He must do some grocery shopping, Neela is coming back tomorrow, and there is almost nothing in the refrigerator.

The next two matches are all scheduled back to back on Sunday. Neela's flight is supposed to be landing around 2:30 P.M. and Krisch's second game starts at 2:45 P.M. Poor Krisch must go to Brian again and Suman will have to miss this crucial game. If the boys win this match they will be top of their league and will be playing quarterfinal next week in Hendersonville, which is about 30 miles north and east of Brentwood.

It is already 9:45 in the morning, and Krisch must go to the field by 10:00 A.M., breakfast is not done yet, Suman tries to find him in his room, he is not there.

"Where is he? Chhutkai (Baba and Ma call him sometimes by that nickname), where are you?"

"I am in the garage, Baba, getting my stuff in the car, we don't have the time to have breakfast, I will just have a banana; you didn't wake up early enough, let's go."

Suman felt terribly guilty; this would not have happened if Neela were at home. He changed his clothes in two minutes and got his car keys. "Okay, let's go; we will pick up some quick breakfast burritos and juice from McDonald's on our way to the field."

All the boys were already on the side field and waiting for their scheduled field to be cleared by the teams that were playing their match. Greetings were exchanged casually, and no one seems to be in a very talking mood this morning, not even Brian's mother. Parents were busy jockeying for their positions to place their folding chairs. This is something very special in all soccer matches; most of the attendees are mothers—

soccer moms; for some reason, it has become a mom's ardent business to cheer and attend their boys' games. Dads don't frequently come unlike Suman. He must come to support Krisch every game he plays. Krisch doesn't mind if his mother is absent but must have to have Dad watching from the sideline.

Today, there is a slight change in the positions the boys are playing, Pat is playing as center-forward and Krisch has been put on the right-wing position. Suman looks at his son as he was positioning himself, no sign of any worries or disturbances on his face or body language as far as Suman can notice.

Chrissie was standing next to Suman, she also did not bring her usual yellow chair today. "Hey, did you notice Krisch is not playing his usual position?"

"Yes, I did, he will be alright." A quick, short answer.

Suman did not want to get into any long exchange on this matter. Game is just about to begin.

"Goal, goal." Krisch just scored the first goal within two minutes of the match. It happened like this—at the beginning Pat, the center forward today, passed it back to the right half-back, Brian, and Brian dribbled the approaching player and passed straight to Krisch, who just ran fastest with the ball before any defender could come up to him and simply placed the ball to the left side of the goalie, and he had no place to stop it. Brentwood moms started to congratulate Suman. They were playing Memphis Rebels, and the Rebels were no match for the Brentwood boys that day. Right from the beginning their morale got demolished with a fastest goal by

Krisch. They lost 3-0. The other two goals came from Patrick's strong free kick and Joe's header from a corner kick that Brian placed so well. Time was 12:45. Suman must go to the airport to pick up Neela.

"See you guys for the next match, I have to go to the airport to pick up Krisch's mother."

The plan was Suman will pick up Neela and come straight to the field. But there was some other plans that were not immediately conceived by Suman's restless mind. He was at the airport around 2:30 P.M., flight is still 20 minutes away for landing. He is pacing and going back and forth to the board where they update the status of incoming and outgoing flights at the Nashville airport. It still shows that the Southwest flight from Denver arrives at 2:50 P.M. No sign of any passenger coming out that he can see from the place he is standing and waiting for his wife to emerge. He then goes to the status board again.

Oh my gosh, the flight is DELAYED, and it is not coming before 4:30 P.M. What is he going to do? Should he go back and come back to the airport? He can't decide. He will miss the last game for this round, next round will be the elimination round and that is the quarterfinal. How is Chhutka going to play? He is going to look back at the sidelines and not find his Baba, will he be able to score? What if he gets hurt? All these questions jam his mind and he feels numb.

Suddenly his cell phone rang. "Hey, Baba, are you coming, has Ma arrived? Let me talk to her."

"No, she has not arrived, and her flight is delayed for

an hour, I think I have to miss your game."

"Don't worry, I will be fine, don't rush, pick up Ma, I will have Brian's mom drop me at the house, I can enter using the keypad on the garage."

"Okay, son, play well."

"I will."

Brian's mom dropped Krisch right after the game. They won the match, 4-2. Krisch scored two goals. Dinner table was relatively quiet. Neela is at home after four days, she missed all the excitements of being there watching her son scoring goals and she is jumping and high-fiving with other moms.

"Why are you so quiet, Baba?" she asks Krisch.

No response for about a minute, and then Krisch says, "Well, I have a test tomorrow and I have not studied at all. I played well, Ma, and scored two goals. We are at the top of the bracket and we play quarterfinal next Saturday, okay!"

End of all small talks. He disappears into his room. The parents exchange glances between them and go about their business of cleaning and other small errands.

Wednesday's practice was short, but the lectures from the coach was relatively long. Boys have won all their five matches and now stand at the top of their bracket for the quarterfinal. Both the quarterfinal and the Semifinal will take place back-to-back on Saturday and Sunday, next week at Hendersonville. Brentwood boys, Krisch's team will play against one of the toughest teams in the conference, Henderson Bears, who was last year's division 2 champion and because of that they moved this year to division 1. So Mr. Wisermann's lecture to the boys extended well over an hour including all tips and strong principled guidance for the remainder of the week, which includes drinking lots of water and no soda at all, eat two eggs in the morning and at least one banana and two glasses of milk every day.

Two days passed quickly and today is Saturday. The quarterfinal is scheduled at 10:30 A.M. and it takes about an hour to reach Hendersonville. Both the parents will be there and Neela in particular will be cheering her heart out for Krisch to score goals. The family was fully ready, lunch and snacks were packed along with lots of drinks in ice chests and few bananas were placed nicely at the top of the cooler wrapped in paper-towels. Back in those days, neither WhatsApp, nor Facetime was available, hence the big brother had to be left alone at his dorm in Chicago. Last night the two brothers talked for hours and Suman could overhear some interesting conversation and tips from the mentor who never really played any soccer, but an avid fan of it and the most biased supporter of his bro. His instruction to Baba was "You must call me Baba when Bhai scores goals."

"I will" was a short answer from the dad.

By the time they reached the soccer complex in Sumner County, time was 9:45 A.M. Finding a parking spot was a challenge. There will be at least sixty games for all different ages and divisions. The complex has twelve fields and there was not even a spot for parking within any close proximity to any of the fields. Reverse parking is not really a Suman thing. Neela always asks him whether she should do it and Suman's ego gets hurt by that. No such requests were offered by Neela, everyone wants to get to Field No. 7 where the Brentwood Eagles will be playing Henderson Bears.

"Just park anywhere, Baba, I must go."

"Okay, I am parking; Ma and I are coming, you better get going."

Krisch jumps out of the car with his soccer bag, and before he runs, comes back and gives his mother a bear-hug.

"Go get 'em, son, and play your best," a simple advice from Suman. "Let's take the small cooler and the two chairs, no need to take the bigger one now; this has four bottles of water, the team manager, Corbin's mother, today is in charge of bringing water for all the players and oranges cut and peeled for the players."

"Hi, Neela, you are back today; when did you get back from your trip?" That was Brian's mother.

Initial greetings were short, and parents are trying to find positions for their placements—chairs and whatever, they brought from home. Neela pointed some place right of Brian's mother and people are about to leave as the match before has just ended, and the Brentwood boys are getting ready to come in.

Krisch is again positioned for right wing, and Pat is positioned for center-forward. This repositioning started during the last game and the coach was very successful, and so was Krisch, he scored two goals. Suman heard last night that he was talking to the big brother about this and he was advised not to make any big deal about it, on the contrary, make the best use of it. Krisch was heard saying that this puts him in a better advantage to go in from the wing and attempt his bicycle kick when there is a good corner kick or an incoming lofted pass.

The toss went in favor of the Hendersonville Bears and they are just about to start. Suman can't sit quiet; he leaves his seat and walks past another parent and goes to the far end of the right corner from where he can see every movement of his son. The ball is being dribbled by Brian from the midfield and Krisch is going zig-zig, closer to the sideline and then back to almost middle and Pat is also running for the ball. Brian passes to Krisch, he dribbles the oncoming defender and passes straight to Pat's stretched right leg. Pat picks up and takes a shot to the goal. The crowd was absolutely quiet.

"Oh, gosh," someone repents, and the shot goes over the crossbar. No goal. This is the epiphany that Suman always had about his boys—*something will go wrong at the right moment.* And time and again he is right, he always tries to tell himself not to manifest any negatives. But mind works on its own; controlling such emotions and thoughts are beyond ordinary people. While he has been totally engrossed with his own thoughts, something just happened, the Bears snuck in a goal. It was totally Corbin's fault; instead of kicking the goal kick to the forwards, he tries to pass it to Joe to the right who was placed as the left back. Joe didn't notice that there was a bear eagerly awaiting; and Corbin's pass was not really dead-close to Joe. So this bear gets the ball and runs past approaching Corbin and gently puts the ball inside the net. The score is now 0-1, Brentwood boys.

Unlike football, soccer games don't have time-outs. So there was no admonishing or a huddle up of the players with the coaches. Brian comes up to Krisch and they call Pat and Krisch just before they go to the center line and start back again. Pat passes back the ball to Brian and he dribbles two players and passes back to Pat, he is about to take the shot to the goal, instead, he passes back to oncoming Brian and Brian extends the ball to the right wing to Krisch, Krisch uses his famous dribbling trick and goes past the full-back and takes a shot to the far left side of the goal post, and it does its magic, it takes the inside edge of the post and goes in.

"GOAL, GOAL," the Brentwood side of the sidelines goes into a full roar. The score is now tied at 1-1.

Next fifteen minutes were just back and forth. Defenders from both sides were on one-on-one and neither Krisch, nor Brian or Pat or Krisch could make much inroads to the goal. And from the Brentwood Eagles, Corbin made up his mistake by saving two goals. It is now halftime and Coach Wisermann corralled everyone away from their parents on to the far-left side of the goal posts so no diversion can creep in the boys' minds.

A ten-minute break seemed like an hour. At the very last minute of the break, boys got some drinks and some of them took a bite of the peeled oranges. They are ready to face the next half and must score goals. Neela and Suman are back now side-by-side seated in their soccer chairs. Neela was unusually quiet today, normally, she is very cheerful, runs up and down and every time Krisch takes a kick to the goal or approaches the goalie, her voice goes up a notch. Today, she hasn't left her seat for a minute. Suman left her alone. He knows that something must be bothering her and better to leave her alone.

"Oh, the game started," Neela startled Suman.

"Go, Krisch, go, go, Brian, go," that would be Brian's mother.

Pat is moving fast to the left making room for Krisch and Brian approaching through the center of the field, Brian passes to Krisch and Krisch runs with the ball, Pat is right behind. Two backs and the defender, no one else in front. Brentwood crowd are all at the edge of their seats and many have their fingers crossed.

"Goal, GOOOOOOOOOAL."

A trance is just pierced through, and Suman pumps his fist while the boys in the middle of the field lifted up Krisch. It happened like this: As Brian and Krisch moving side by side and one passing the other and Pat right behind, Krisch back passes the ball to Pat and instead of taking the shot to the goal, which would have been difficult since it was at an angle and about 30 yards far, Pat passes to Brian and Brian tossed the ball high enough for incoming Krisch to jump and placed his head on the right spot of the ball and it went straight into the net. Not a bicycle kick but an awesome header. The tournament favorites are back in the game. Rest of the game went very tough, some of the Brentwood boys got hurt, Krisch got hurt, so did Pat. But Pat is a tough boy, he can handle it, however, Coach must replace Krisch from his right forward position and bring a substitute player for the rest of the matches, which is a semifinal and if they win then there will be the final. The final whistle came down after a three-minute extra-time play, and the boys fought very hard to protect their turf and Corbin rose up to the occasion and did not concede any goal. The final score remain at 2-1, Brentwood. Next game, the big semi, is at 2:30 P.M.

Chapter 4

The Semifinal and the Bicycle Kick

Congratulations are in order, everyone from the parents' club started to congratulate the boys, particularly Krisch for his brilliant and timely header for the win. Even Corbin's mother came up to him and gave him a hug.

"We need to win the next game, Krisch, play your best."

Krisch just gave a nod. He was very hungry and wants to lie down at the back of their Toyota 4-Runner and eat the food, his mom so neatly packed. Lunch was simple, chicken salad sandwiches, chips, chocolate chip cookies, and water. No soda at all. Suman wanted to have some Coke or Pepsi, but didn't even mention it.

"Well, we have almost an hour and a half. Why don't you take a nap and relax those feet and your knees, we will turn the music in a soft tone and try to get fresh, I will wake you up at 2:00 P.M., a half-hour before the game."

The semifinal is against a team from Knoxville, TN, they are last year's regional champion for the Southeast. Krisch did not play last year since he just moved from Kentucky. Brentwood boys played them in Memphis at quarter final and lost. Today is a repeat match and this time Brentwood boys must win. Six of the eleven players are new and not faced these Knoxville Ruins. And Brentwood boys, except Corbin, Brian and Pat and Krisch do not have much reads on this rival. Coach, as usual takes his team to the farthest end of Field No. Nine, where the game is just about to begin. Intense discussions were going on without any parent nearby and the assistant coach with a clipboard showing the boys their relative positions and passing, taking shots at the goal, etc. First whistle was heard, and the linesmen are on the ground. They all just did their usual routine with high-fives at the end and parents start clapping with the cheers—"Go, Brentwood, go, go, Brentwood Eagles, go."

Brentwood won the toss and decided to take the north side of the field. Today's captain is Brian. He calls his team on the sideline and had a word of encouragement, he then goes near Krisch and says a few words. Both gave a thumbs-up sign and off they go to the center. The center forward for the Ruins is a stocky short guy, perhaps five feet one, he looks daunting and passes the ball short to his right half-back and moves himself near him and starts advancing, Brian comes to tackle him.

"Oh, my, Jessy!" He passes ten feet to the left of Brian and runs fast to pick it up, exactly same technique that

Krisch uses many times. He picks it up and then faces two backs guarding him and a third Brentwood boy guarding the nearest player. Corbin, being the best goalie has a pretty bad tendency, he advances, leaves the goal unattended and anyone, if runs past him has an open net. This time Roberto, the Ruins center forward, passes the ball to his left wing who was advancing fast; however, the pass was incomplete and went out of bounce, outside the sideline. Throw-in for the Brentwood. A long throw straight on to Krisch's extended right leg, no one was in front of him, Krisch runs with the ball—two defenders come and charge Krisch, and it was relatively rough. Whistle blows: FOUL, referee comes running. A yellow card for the one who targeted Krisch's right thigh but could not reach there quite, but got the outside of the back of the thigh: Krisch got slightly hurt, but brushes it aside. A free kick. Brian comes to take the shot, asks his forwards to go all the way up to the goal. A beautiful kick and Pat takes a strong header but straight into the hands of the goalie. No goal.

"Good attempt, a very good attempt." Suman just looked at Pat's mother. She acknowledged with a smile.

The boys gather themselves quickly and this time there is a change in their strategy; two of the four backs are moving parallelly with the half-backs and all are in the offensive, leaving some gaps. Suman points this out to one of the dads who also noticed that.

"They must have some plans, let's see."

Suman goes back to his self and didn't say a word anymore. He knows that the Americans don't like to be too critical of themselves, especially when it comes to sports, and that too soccer which is not a mainstream sport yet. Brian is leading the charge; he passes to Pat and all the four forwards are circling in and out as if they are forming a bowl of fire and then going to erupt out. All on a sudden Suman finds that Krisch is not in that formation, he was not even in his position, he is way to the right of the typical position that he likes to play—right wing, but closer to the center. This time he is way closer to the sideline, as if just about to take a throw-in. Now, there he is; comes charging, running fast, passing one lone defender trying to guard him, the ball is in Brian's left foot, he lobs the ball just about waist high, too low for a header and Krisch is still approaching, he is not quite there yet to take any shot or a header, and the ball is only about ten yards from the goal.

"Goal, goal, GOOOOOOOOOAL." The Brentwood side of the sideline bursts into loud and thunderous roars. Neela is jumping in joy, all the dads high-fiving Suman—what a scene.

This is how it happened. As Brian lofted the ball, waist high about five feet ahead of Krisch, he could neither take a shot with his extended right foot, nor can he take a header, he just does the impossible, he lungs himself to the back of the falling ball and put his left palm on to the ground and takes that *bicycle kick*, and what a motion that was! all synchronized and perfectly timed; the foot, the angle, and the body all in unison, as if imitating the art of Da Vinci; or the baton of a Maestro asking his full ensemble to play the finest of the fine tunes in F-major. Suman shook his head and tears began to roll down from his eyes in joy. Neela was right next to him and squeezed his hand. He got hold of himself.

"What a goal, what a goal!"

Postscript: Rest of the game was placid, just up and down. The score ended 1-0, Brentwood.

Final: Brentwood lost 2-0 and got the Runners cup.

The Ball Drops at the Big Apple

Chapter 1
Planning for the Trip

"How about New York, Baba? Let's go to New York and spend the New Year's Eve there, it will be the end of a millennium and the beginning of another."

This was the first sentence Soumit said at the dinner table helping to set up the table and obliging finally Mom's repeated requests for setting up the table. Well, none of the boys liked their recent move from a comfortable and familiar place in the old Southwest town—Las Cruces, where Soumit spent his past twelve years and Krisch his first ten years since birth. They were much less talkative at this new place, and to their parents' dismay, not fighting at all. The younger one, Krisch, was always little introvert, he talks to himself when left alone, finds various pieces of weird things around the house and if asked says, "I am working on these." What? How? No one knows. His brother used to tease him all the time and occasionally got lots of scolding for this from both his parents. Today is different. The move took a big toll on the family and the imminent split is pinching everyone's nerves.

American universities typically do not hire their own graduates, unless it is Harvard, or MIT or Princeton. They want you to go out in the wider world and prove your worth. When Neela defended her Ph.D. dissertation, Dr. Blake, one of her committee members, told her, "Girl, you need to spread your wings now. Go find your call, wherever that be." Ever since that day Neela had been trying. After six campus interviews, she picked one in Kentucky, that too a small Catholic university. She felt very good at the interviews that she had when she visited the campus; new program, she must develop a master's degree program in her area—Special Education with emphasis in Behavior Disorder. Salary was not great, but manageable. The idea was that Soumit will stay back in Cruces with Baba and finish his senior year and Krisch would accompany the mother. Baba or Suman will continue as is and will visit the family as much as possible. However, there were some other plans in the making, quite unfathomed.

On a cold day, Las Cruces rarely gets too cold, the family packed their Toyota 4-Runner which Suman just bought for Neela since she will be staying in a relatively much colder place and must drive in snow and they were just getting ready to drive 1260 miles to their split home, Owensboro, Kentucky.

"Well, we are just about ready, Krisch, why don't you and Dada sit at the back since you guys are not quite awake yet, Ma will sit in the front and no shotgun, please. Just relax and don't fight, we have a long journey ahead of us. Our plan is to drive till Oklahoma City today, that's about 660 miles and then tomorrow we will do the rest."

"Okay, Baba," Krisch's quick but slightly morose response.

Soumit, on the other hand, said nothing, he just picked up his violin and took his sit at the back of the SUV. He started strumming the violin to the famous tune of "Gavotte" by Bach.

Chapter 2

The Journey Begins

The family started their long ride to East. Owensboro is a small semi-urban town with about fifty-five thousand people living then. The apartment Neela selected after talking to a few realtors is conveniently located, little off from the main road and about equidistant from her University and the Daviess county middle school. A two-bedroom with a kitchen and a small living area, about 1400 square foot just enough for the mom and the younger son (no one knew then that the older boy will move in, in a matter of days) and two separate baths. Rent was modest, Neela can handle that on her own without any support from Suman. That was a discomfort for Suman in the beginning, but he had no other way but to accept this decision of Neela since they will have to maintain two establishments and hence a bigger apartment or a house will be out of the question, at least for the time being. It was a semi-furnished apartment, few things can be purchased slowly, that was Neela's quick reply to the boys when one of the boys was asking for another couch, or another bed in the second room. This was all arranged from a distance, Neela did all her arrangements only seeing some pictures sent by the realtor via internet, well, it will only be for a few months until they find a better place.

"Hey, Dada, can you get those boxes from the back of the SUV, they are too heavy for me."

Soumit was not listening to his brother's calls, he was still strumming his violin with his long stretched right leg. Suman comes up from inside of the house and speaks to his younger one, they both went inside and brought back the boxes.

It was December 19, 1999. In a few days, the family will drive to Indianapolis and spend the Christmas Day with Suman's cousin brother's family and then on the 26th they will fly to New York. The car will get a break from a long journey and rest at the driveway of the cousin's brother, and after the New Year when the family is back from their millennium celebration will pick it up and drive to Owensboro in search of their (at least for half of the household) new life.

"We are all packed, Ma."

"Why don't you come and take your seat and let's get going?"

"Vamonos, Amigos."

Krisch is taking his second semester of Spanish at sixth grade and hating it every moment. But he will surprise everyone speaking Spanish occasionally in perfect grammar and phonetically correct way. Soumit is something different, he hated Spanish, after being in the class for one day, changed to French, and stuck to that for three semesters, that ended last year with an A in the class after baking a French-vanilla ice cream cake at home and presented the whole process in front of a class jury. His French teacher was a German-descent woman in mid-forties and told Soumit's mother that she has never met anyone like Soumit's charming personality. Mom felt pride in that but never shows that in front of her son, that was not appropriate for an Indian mother.

"Hey, you all come inside for a minute, we need to do our prayer and pranams before we proceed." Neela's voice was soft but firm.

All went inside and Soumit finally came slowly with his head slightly down. He could never accept the idea that the family is going to be broken and he has to stay with his father to finish his high school, three more semesters. But never grumbled much for he knew how much it meant for his mother, finally been offered a tenure-track job at a four-year-plus university.

Prayer was simple, Neela did the chanting of their Gyatri mantra:

"Om Bhur Bhuva Swaha Tat Savitur Varenyam, Bhargo Devasya Dhimahi Dhiyo Yonah Prachodayat" (three times, meaning let the whole world and my world be one and all in unison and peaceful and all be blessed by the Almighty, be with us and give us strength to overcome all sufferings…). Slowly they all went back to their designated seats, the dogs were already taken to a farmhouse in Messilla and the owner agreed to keep them for two weeks until Suman and Soumit get back.

Their plan was to leave Cruces by 8:30, they left around 9:30 A.M. The first stop was around 1:00 P.M. at a Burger King along the I-10 East. Food was simple: two Spicy Chicken Burgers and two Classics, boys cannot eat spicy yet. Getting back in the car led to some commotion, this time the younger one called shotgun and sat next to Baba, older one got terribly upset, arguments went on for about three minutes, Ma had to intervene and calm them down with a compromise that there will be no shotgun next time, older one gets to sit in the front.

Driving through Indianapolis could be challenging at times. Every time Suman comes to his cousin's place they always find ongoing construction.

"Can't they take care of these roads once and for all? At least for the next ten years?" Not a question, thrown to anyone, almost a mumbling soliloquy. Soumit, was wide awake and made a profound comment: "How will the contractors and all their workers survive, Baba, if they complete it to the specs? They must leave something incomplete so that they can come back later on a different contract, perhaps a bigger one, and that's how the contracting cycle goes on." He was in 10th grade, had lots of AP courses, including Economics, American History and Political Science, besides AP math and physics. This comment of his made Suman thinking of developing a model, he later called it "Fix-it-Later" principle in public economics and will get several accolades in his profession, he will also bestow full credit to his older son for this notion, now that is a different story.

Chapter 3

Christmas Eve at Cousin's

They all reached Suman's cousin, Shankar's, place around 8:30 P.M. That was Christmas Eve. Everyone at Shankar's house, his wife and the daughter, Sneha, were all waiting for their relatives to come and join them for a Christmas Eve dinner and then watch some TV in their family den till midnight, and then open their presents before going to bed. Dinner was simple, chicken curry and rice and some potato fries, ice cream for dessert. Dinner or lunch is always simple at Shankar's house, for his wife, Mani, does not like to cook, and Shankar ends up doing all the cooking. Neela used to tell this to Suman but Suman never cared for these sort of things and told her many times that "It is their affair, let's not poke our noses into it." Neela obliged and doesn't grumble anymore, if she could, she would have brought some fried Ilish (Hilsha) fishes for her cousin brother-in-law, but she had to finish lots of schoolwork as well as submit second article from her dissertation for publication. She knew that once she starts teaching in January and building the master's program she will not have much time to publish, at least for the first year.

"Hey, guys, kids, dinner is ready, why don't you three sit on the table, adults will eat afterwards," Shankar, nicknamed Shanku, announced. "Once everyone finished eating, we will have a music time, we will hear violin from Borda (Sneha calls Soumit Borda), cello from Krisch, and then keyboard from Sneha, and if Boudi can sing one or two songs that will be great."

"Sounds like a plan," Neela quipped.

Unwrapping the gifts and then having a "WOW" or an "okay" moment has its own place. Neela chose three gifts for the family and wrapped them very carefully with matching bows and placed them under the tree within minutes of their arrival. She noticed there were three boxes wrapped without detail attention and lying haphazardly, as if someone unwittingly or carelessly put them there to be brutally torn and get to the bone hurriedly. Not a pleasant feeling.

It was Christmas Eve, so there has to be some Christmas Carols, "Joy to the World" is for everyone. So the whole assembly sang that, followed by "Silent Night."

"Now it is time for the Maestro, play your favorite piece, Soumit," that was a request from Chhotokaka, the boys called Shanku, Chhotokaka. Soumit was learning Mendelssohn at that time, not quite complete, however, learned enough to perform at a small gathering. So he got ready and fetched the violin from inside the room where the brothers will be spending that night. He played his heart out, tears were flowing from his mom's eyes, at least Suman could see it.

"Bravo, bravo!"

Now it is the little Maestro's turn.

"What are you going to play, Krisch?"

Krisch looks at his mother and then to his brother. Big brother must approve anything he plays, or for that matter anything he does. Soumit has a supreme influence on Krisch. After a couple of exchanges of looks, Krisch played "Gavotte" in G-minor, it was alright, not eloquent like the older one.

Big brother cuts in, "Hey, Bro, let's play Bach Double, shall we?"

Little hesitation from Krisch, but he cannot refuse his older sibling. So they played. Again everyone cheered, "Bravo, bravo." Clock almost struck twelve. Few more minutes, it is Christmas and time to unwrap the presents.

Both the boys got sweatshirts from their Chhotokaka. Time to open the presents for the hosts. Neela gives lots of thoughts to the details, she wrapped a cashmere scarf for Mani, a polo shirt for Shanku and a green velvet dress for their three-year-old Sneha.

They all praised their Boudi (older cousin/brother's wife is called a Boudi as a show of respect) and Suman and Soumit exchanged glances in appreciation of their wife and mother respectively.

"Boys, please go to sleep, we have a flight to catch in the morning." Their flight from Indianapolis was scheduled to leave at 11:30 A.M. and it will take about 45 minutes from Shank's house to go to the airport. As per the plan, they will leave their Toyota 4-Runner with Shanku and retrieve it back on Jan. 1, when they get back from New York. Neela must start her semester on the 3rd of January and Krisch's school also starts the same day. Soumit and Suman will go back to Cruces on the 4th. Such was the plan, however, often times there remains a big gap between the plan and the reality.

The flight to New York was smooth, even though it was December, weather was not unfriendly either in Indianapolis or in New York, temperature was in low twenties, not unbearable. Their stay was at a friend's apartment on 48th Street, a small one-bedroom apartment with an exorbitant rent. But Sunil could afford it, he was a budding Wall Street investment banker. He has gone to India to visit his ailing mother, hence Vikram, Sunil's best friend, was staying there for a couple of weeks and Vikram was the adopted younger brother to Suman and Neela, and was very fond of his adopted nephews. Vikram was waiting at the gate to greet the family. After initial hugs, and collections of the luggage he led everyone to the exit.

People don't usually drive in New York. Ask a New Yorker, and you will hear a sense of pride in their voices when they will say, "I hate driving," and then finding a parking is impossible, so use the taxi or buses or the metro. Vikram took a minivan taxi to Sunil's apartment.

New York apartments are small but efficient, this old city has offered its charm throughout the past three hundred plus years and still continue to attract new settlers, be it a young banker, an artist, a techie or just a freelancer. Sunil came to New York, Manhattan in particular, from his previous job in San Francisco, where he was an investment banker at Wells Fargo. This new job at a top Wall Street investment bank affords him a bigger personal (financial) growth. He was planning to get married before he turns thirty next year and buy a house in Connecticut. So this apartment was a good choice for a while.

"Ma, where Bhai and I going to sleep?" Soumit asked, looking at the small place.

Vikram was nearby, he immediately picked up and answered, "We, three of us, will sleep on the floor on sleeping bags, it will be alright for a few days, we will all have fun, won't we?"

The younger one immediately said, "It's alright."

The arrangement was – Neela and Suman sleep in the bedroom and the boys and Vikram sleep in the living room on the floor. There was still enough room to place the sofa in a diagonal way for another person to sleep on it, and that will be Sunita, who will be flying in, day after tomorrow on the 29th from Albany. She is a common friend to all and used to live in El Paso with her ex-husband Frank until their divorce. She moved to Iowa after their divorce and got her Ph.D. in Sociology and now is an Assistant professor in SUNY, Paltz. Boys did not like the idea that there will be another person in that small place, that too a woman. But younger folks (especially teenagers and preteens) do not have much say in Indian households. Settled as they may, all planned to go out for a lunch and Neela wanted to buy some grocery since eating out every day is too expensive, and her husband has a strong affinity for home-cooked food.

Choices: lunch, dinner, breakfast or snacks and in between in New York have abundant presence. Vikram offered Indian, boys were not too excited, they preferred Italian. So the contingent had to walk a few blocks toward 48th Street and situated at Pablo's Grand Italiano. Neela only selected a chicken Caesar salad with olives and feta cheese while Suman chose a slice of pizza and minestrone soup; the brothers ordered meatball pasta and Vikram did the same. When the waiter brought back the check, Suman wanted to extend his Visa card, but the waiter only gave a receipt to Vikram and said everything is taken care of. Vikram already paid. "You shouldn't have done that, Vikram," Neela complained. A return smile of gratitude, all she got from Vikram.

The next day was a whole day trip: first to the Ellis Island, Statue of Liberty and then to Broadway. As the ferry was approaching the Island, cool breeze from the east was making the younger boy shiver, he came close to Suman and was trying to get as much warmth as possible. The older one, just turned sixteen two months back, was wearing his new Columbia jacket (birthday present from Mom) and Doc Marten boots and was enjoying the breeze. He and Vikram were chattering nonstop on topics that transcend from current politics to science and to music. Neela was slightly cold and got closer to her husband and the young one. Suman has this seasickness that has been developed in the Caribbean when he worked there for the United Nations, he remembers, one time he was in a yacht with a Canadian older couple who loved Neela and him as their younger friends. Suman had a beer and within a few minutes he started throwing up, what a mess. He felt so embarrassed, but John and his wife, Betty, comforted him as if it was nothing. Since then he has been careful, but it happened over and over. So he did not eat anything that morning, when asked by his boys, his answer was short, "I will eat at the cafeteria there—at the Statue of Liberty Park."

The park was totally full, after they bought the tickets, this time Suman paid for all five; must wait in line for at least an hour and a half to get inside, and then it will be at least another two hours going up in a single line one by one. No one complained. Boys are excited. Even though the older one visited once before, but he was only four years old at that time and the younger one was not born. It will be Vikram's first time also. So the uncle (even though they are not blood related, it is customary in Indian culture to call an adult male uncle) and the older nephew delved into their discussions while the younger one gets closer to his mom to get little more warmth. Suman noticed that Soumit was not paying much attention to the details as they were all climbing the stairs, his mind was somewhere else. Suman murmured the feelings to his wife and Neela said the same thing, she added, "Since last night, I noticed that he was very indifferent, this happened after he was talking to someone on the phone."

"Who was that?" Suman asked.

"I think it was Pia from Chicago."

"Did you hear anything about their conversation?"

"Please move."

"Oh, sorry."

People in front of them have climbed up a few more steps, all of a sudden there is quite a gap. Vikram and the boys are waving at them. Somewhat embarrassed, Suman and Neela scaled the steps quickly. They both changed the subject and got immersed into their immediate future of two separate living arrangements, split between Owensboro, Kentucky, and Las Cruces, New Mexico.

"You know I am not going to buy any new furniture at the time being."

"Well, Neela, you have to buy at least one futon bed and a dining table. What about the beds?"

"I will rent them, there is a store in Evansville, they rent all furniture by the month. We will go there after we return."

"Okay."

"When do you go back?" Neela looks at Suman.

"On the 3rd."

Neela heard a deep sigh in Suman's voice. Neela knows that this temporary separation is tearing up everyone, especially Suman. He is very close to his younger son, and leaving without him is a dread for him. But for Neela's academic career and family's growth he halfheartedly accepted, what else can he do? He was in no position to get Neela an university job at his own.

A 15-minutes stop at the top or at the base of the Crown was all they had bought the tickets for. Going up the Crown was not a possibility since all tickets (a very limited numbers are only issued each year) were sold out way in advance. Everyone was very excited to be at that vantage point. From the other side of the Manhattan, everything looked so tiny, even the Twin Towers (yes, that iconic New York symbol) that were still standing by as if they were defying the gravity and the ruthless meanness of today's hatred, bigotry and racism.

Climbing down was relatively quicker. It only took them twenty minutes to come down.

"Are you all hungry? I am starving. Let's find the cafeteria, it is on the back side of the entrance," Vikram said, and there are restrooms.

Wondering on the Broadway was not in the agenda. Something changed while Vikram and Soumit were talking on the top of the roof, just below the Crown. The plan got totally changed. Originally they were planning to visit the MET, but Vikram suggested that the four of them do that the next day while he will go to the airport to pick up Sunita (a common friend) who is supposed to arrive from Albany and spend the next two days with them in the same apartment. Sunita just got divorced and has been going through an emotional trauma, needed to spend a few days in a different settings without the constant and nagging feeling of missing Frank, her ex-husband. But that was not the main reason for Vikram and Soumit to divert everyone's attention and careen them on to Broadway.

Pia or Piyali and Soumit came to know each other about two years ago at a Bengali Cultural festival in Santa Clara. Soumit, as usual, mesmerized everyone playing a few western pieces, Mendelssohn in particular and then for the finale, played "Purano Sei, Diner Katha" by Tagore on his violin during the youth sessions. Many young boys and girls came to shake his hand, and Piyali was one of them. Since then they have been talking over the phone at all hours whenever, each found some time. Pia invited Soumit to visit them at their Chicago residence, Soumit's mother was not very excited about it and so far that did not happen. Sonia, Pia's mother, on the other hand was very supportive of her daughter's overtures for this charming violin player, and also a topper in his class. No one knew that there was some plan in the offing. Soumit confided something to uncle Vikram, hence the whole gang was on Broadway that afternoon, even after a tiresome all-day excursion to the Ellis Island. Time was about 5:30 P.M., time to get back to the apartment and arrange for dinner, Neela did not want to spend any more money and just cook some Kitchuri and Begun vaja (fried eggplant) for dinner and may buy some Kentucky Fry for the boys. That plan had no place for happening soon. No one noticed that they are walking by the famous Plaza Hotel on the Fifth Avenue. Vikram and Soumit at the front, about fifty feet ahead of the rest of the gang. Suddenly, Vikram looks back and waves at Neela. Neela looks at her husband, asking—what is happening without uttering a single word. Suman looks at his younger son, he just shrugs! They all tried to speed up to catch up to Vikram and Soumit, but Vikram was walking back. He just says, "Just watch!"

It was almost six in the afternoon; sun has just started to retire for the day and long shadows of the tall edifices started to create a hallucinating environment on the streets of New York. Soumit was standing tall looking at his left and another figure just emerged from the left, a shorter version, of course a female. "Oh, she is here!" Neela exclaimed.

"Who?"

"It is Pia," little bro just blurted out very nonchalantly.

"Hey, did you know this?" Suman asks his younger one.

"No, but I guessed," was his very casual response.

"Vikram, did he tell you about this, and that's why you guys changed plans, right?" Neela asks Vikram.

Vikram just laughed out loud. They were all walking while surprised and found themselves in front of the entrance hall of the Plaza Hotel. Pia's mother, Sonia, was standing there with her daughter and Soumit.

"Hi, I am Sonia, we just arrived here this afternoon, come inside, let's have a coffee, and then we can go out for a dinner."

Neela was slightly taken aback, she gave a quick look at her son, who gently looked back at her mother and then lowered his gaze.

Vikram broke the ice, "Hi, I am Vikram, you must be Pia, Soumit told me all about you when we were at the Statue of Liberty today. I am dying for a cup of coffee, let's go to Zeni's. I know this fabulous place, it's only two minutes' walk down and we are there."

Zeni's Gourmet Coffee is not an upstart coffee house, not like today's Starbucks. It is an old specialty coffee shop with a mom-and-pop look—old sketches of New York, London, and Paris, a few motifs (printed version of course) of Rene, Warhol, Rembrandt and Picasso. Sitting arrangements had to be rearranged, for seven people they had to put two tables joined together; at the far end Soumit and across from him Pia sat, she was bubbly, and her mother was eyeing Soumit like she wants to get them engaged right there and then. Neela was observing and was trying to keep her cool. Suman was trying to a do a few small talks with the newly re-introduced (they met two years back at Banga Sammelan in San Francisco) mother and daughter. Sonia ordered coffee for the adults and hot chocolate for the teenagers and the preteen with some almond cookies and cashew nuts.

"Why don't we go to this Indian place, called Bombay Palace?" asked Suman.

"It will be impossible to get any seat tonight, they get all booked from weeks in advance, I know a Thai place, not too far, about four blocks from here, and we can possibly reserve seats now."

Neela just gave Vikram a smile. Soumit was not talking much except exchanged a few college search tidbits with Pia. He wants to go to Northwestern, and wants to pursue a double degree there, in Music as well as in pre-med track. Took pre-SAT, and will be taking SAT in March.

"Neela, why don't you come to Chicago next month with Soumit, and we can visit the school, I will call in and arrange a visit there," Sonia said so casually but very sincerely.

Neela was feeling little uncomfortable, but didn't show her traditional emotion, just said, "Okay."

Soumit just looked at his mother.

It was about 8:00 P.M.

New York was bustling with millennium ending glamor and festivities, lights of all color were dazzling the city inviting all to join the party, and everyone, at least that was the feeling— outside. "Well, our dinner appointment is at 8:30, and it will take about 20 minutes to walk down there, shall we go?"

Vikram almost led everyone toward that direction. Neela and Sonia were walking at the tail of the group, chitchatting about their daughter and son and their college choices, Soumit and Pia were walking close to each other side by side, Vikram and Krisch were talking and walking, mostly Suman was walking alone just a few steps ahead of Neela and Sonia.

The Thai Cuisine at Lexington and 54th Street was only about a seven-minute walk from the Plaza Hotel where Sonia and Pia were staying for two nights. But it took the gang about 15 minutes to walk that short distance since they were walking and talking and appreciating the ambience of all New York.

Dinner took about two hours. When the waiters brought the bill, Sonia was trying to get it, but Suman grabbed it and put his American Express on the plate which had the check. Neela just gave him a nod of approval and that Soumit looked at his father with a proud glance. Pia's father is a well-known invasive cardiologist in Chicago, and naturally rich, even though Sonia and the family try to keep a modest and cool social vibe, it automatically shows in their dealings. Flying in from Chicago for only two days just to be with Soumit for a few hours and staying at Plaza for five hundred dollars per night was a bit obvious showoff. Then who knows, infatuation at this tender age has its price.

"So, what's the plan for tomorrow? I have to go to the airport around 2:00 P.M. to pick up Sunita who is arriving from Albany, and then will be free from 4:00 PM. Why don't you guys go to the MET or something."

"That sounds like a plan, Vikram, you are good," Sonia complimented.

"Well, I would like to see a Broadway show, I came here with my choir group ten years back and saw *The Phantom of the Opera* after we sang in the Lincoln Hall, Vikram, can we get tickets for *Cats*?" Neela asked Vikram.

"We can certainly try."

"Well, ticket office is closed now, but they open at 10:00 A.M. for walk-ins, we can come in the morning to check out. Are we all going?" Vikram asks, looking at Sonia.

Everyone, except Krisch, said yes. He wants to go the Science Museum and see how robots work. Younger voices are always ignored, but he doesn't complain, he lets his older brother dictate terms. He knows his days are sure to come. So the plan is like this—Chicago people will meet the rest of the gang around 4:30 P.M. on Broadway, will have early dinner, and then go experience the *Cats*. Vikram called a taxi for Sonia and her daughter. After a few calls from the mother, Pia finally boarded the cab and off they go to the Plaza Hotel.

Sunil's apartment, where the family was staying was not very far, and walking in New York is a pleasure specially with all that is going on tonight.

Cats was mind blowing. Everyone has been awed. Neela was very happy to see it. "Vikram, how did you manage to get the tickets?"

"Well, Sunil has a banker friend who has some inside contacts, I knew him and met a few times here in New York, I called him up last night and he got us these tickets, and that too at a very discounted price."

"I am very impressed," Sonia commends Vikram.

"I am hungry," Krisch who typically does not talk much just announced, and everyone including Pia noticed and says, "We must find a good Italian place." She knows that the brothers love Italian, however, Neela doesn't like it that much, but she can sacrifice her appetite for her boys.

"Little John" has been an icon on the 12th Street in the East Village. Has been serving since the beginning of the 20th century. Locals call it Lil' Johnny. The actual name is John's. "If you want authentic pizza, this will be it, and Neela di, you will also have your choice, trust me."

Vikram took the credit of introducing the Old Lil' Johnny to the crowd. "We need to get two taxis to get there, it is on the lower side of Manhattan and in the Village, the Village is typically known as Little Italy, of course.

Order was simple: two large hand-tossed pizzas with pepperoni, mushroom and olives; one large vegetarian lasagna and Sprite and Diet Coke for Sonia. It took about 30 minutes for them to get the food, and only five minutes to devour them. Only Neela was still nibbling on her plate. Pia and Soumit were still sipping on their drinks, Pia's big eyes were fixated on Soumit's. Suman looked at Sonia and found her enjoying every bit of her daughter's overtures. Neela was nonchalant, and her mind was somewhere else. She was perhaps thinking about her new assignments as soon as they go back; Suman has to go back to Las Cruces, and Krisch will start his new school without his big brother coaching him what to do and what not to, so she was pretty unmindful.

"Shall we go?" Vikram asks.

"Yes, yes, we have an early morning flight to catch, Pia, let's go."

Sonia waves at her daughter who was sitting at the far end of the table holding Soumit's hands. Reluctantly all got up and headed for their respective destinations.

Goodbyes are never pleasant specially when two young teenagers are infatuated with each other. Pia got on the taxi after her mother almost yelled at her. Sunita was home alone, she didn't join the group to see *Cats* for she just saw the show previous week and she said she must finish her grading by tonight, so she stayed back.

"Hi, Sunita, how are you, I hope your gradings are all done, did you eat anything?" Neela's voice was sincere.

"Yes, I did, and Vikram had some leftover from the lunch, I ate that. We can have some hot chocolate; how do you like that, boys?" Sunita tried to be a team player.

"I would love some," Vikram said, "we all will have some."

"Let me help you." Neela veers off to the kitchen, or a kitchen like place, these new York lofts have a kitchenette just to accommodate a single or two single people's daily business.

Sleeping arrangements became little tighter since the addition of Sunita put little extra pressure on Vikram, however, one night will be not unbearable, they will all manage. Tomorrow is the day they are looking forward to. Everyone recommended that they go to Times Square around noon and spend the whole day there (12 hours) just partying and wait for the moment, the passing of the old millennium and the dawn of the new. There will be programs, telecast live all day including MTV shows, performers from all over the world on the street beginning around 5:00 P.M. till the BALL drops and then continuing on to the wee-hours. Excitement all around except Soumit was silent, his unusual self was bothering Suman and Neela, but they left him alone. Vikram made hot chocolate for everyone and it was already 1:00 A.M., time to lie down on whatever best can be arranged in that cramped place.

Sleeping was not that easy, especially for Suman. He and the boys slept upstairs and Neela, Sunita and Vikram lied down on two sleeping bags in the living area. Bathroom was next to the kitchen, so Suman had to come down to take a leak, and it was about 4:00 A.M. He found all the three were still talking, and it was mostly about Sunita's situation. He wanted to stay back but Neela's look made him uncomfortable and he gently retraced his steps back to the loft.

Chapter 4

The Unforgettable Day December 31, 1999

It was around 9:00 A.M., Suman woke up from some sounds coming from downstairs, nothing unusual, just pots and pans, typical of their own concerts when put on to action. He came down to see that Neela and Vikram are getting breakfast ready.

"Oh, Sumanda, you are up! Did you sleep well?"

"Well, I did. Boys are sleeping like logs, let them sleep for another hour, Vikram, what time we should leave and walk towards Times Square?"

"Well, there are some changes in the plan—Sunita is not feeling well, she may have a fever, and I have to finish a report today for my office since I am away for five days already; they called early this morning. So you guys go ahead, whenever, I mean as early as possible. I will draw a map here, how you can get to the Times Square, easy, and then getting back to the apartment, okay!"

At the breakfast table when Vikram told them about the changed plan, the boys looked slightly dejected, their favorite Vikram uncle will not come with them, only Baba and Ma! Not much fun, but neither of them said a word. Vikram did not look at Soumit. Slowly they finished their breakfasts and went back to the loft to get ready. As Neela and Vikram were putting the dishes back to the dishwasher Sunita came down wrapped up in a blanket, she really looked sick. Her voice was faint, and she started apologizing but Neela stopped her with a comforting voice and cooled the situation down.

"We will be fine, guys, don't worry, we will get ready in 30 minutes, time is now 11:30, boys, get ready, we will leave around noon." She started walking up the stairs to get ready.

The walk to Times Square did not take more than 35 minutes. The little direction that Soumit got from Vikram was perfect. Now he is in the lead with his little bro in front of Ma and Baba with the map in his right hand. Crowd has been swelling beyond Suman's imagination; only 12:45 P.M. and looked like jam-packed streets with more than half a million people. They were trying to find a cozy place near the main stage closer to the one-time square building from where the ball drops. Nothing doing. Every inch has been taken, no place to stand even, so they found another place closer to MTV stand where Carson Daly will be doing his EMCEE tonight. It was also closer to a restaurant that they can order food and a relatively easy way out to go to the Johns (Porta Potties) that were placed about 200 yards from the nearest crowd gatherings. Time was now 1:30 P.M. Almost ten and a half hours to spend standing, squatting and possibly leaning against each other's back in New York's sunny but cold afternoon.

Lunch was not an issue, Neela made some chicken sandwiches last night and they had water bottles that Suman carried in his backpack, and Soumit had the newly bought cannon camcorder in his backpack. Crowds were just pouring in. Neela's piercing eyes are always on her older son. She found Soumit, about 40 feet to her left talking to a few couples, and their accents told her that they are not from here. And look, Soumit, is coming back and the two girls are also approaching them. "Look, Ma, this is Carly and she is Meghan and there is Rob, they are from Holland, just came to experience this amazing beginning of the new millennium." Introduction and shaking hands were quick. They all wanted to go back to their circles. "I will be over there, guys, bro, you can come, we will be back here soon." Brother immediately obliged. Time was 3:30 then.

At least three more hours to while away before any cultural programs and floats start plying in that cordoned zone. Neela whispered something in Suman's ears and both were looking for the boys, Suman found one of them, and approached him, said a few words and pointed towards a place little left of where they spent past two and a half hours. And that place was right next to the MTV booth and just below their quick-made broadcasting station for the special event. A big screen was set up right over and everything was set up to stage a big show, now just wait for a few hours. They started playing the popular music and people were feeling much energized. By this time there was not even any place to accommodate any new entrants. Police were cordoning off all entry points to the Times Square, if one needs to leave, there will be seldom any chance to find a way back in. If one needs to go out for a bathroom break, one needs to approach the security guards and get an exit pass, and only with that pass, he or she may come back. The place that Neela found was even closer to the Restaurant, even though the restaurant was not allowing anyone to just use the bathrooms, Neela was able to sweet-talk them and order some snacks and made wise use of the facility. She was also able to take the younger guy with her and he could use the little boys' room. When they came out, the men went outside for the Porta Potty. Someone must be there to keep their squatter's rights at least for next few hours.

Bill Clinton was the President of the United States and Rudy Giuliani was the Mayor. Clinton just delivered a commanding and very reassuring address to the Americans and to the world urging all for getting rid of divisiveness and racial and cultural bigotry, and quoted Dr. King. NBC, and its anchor Tom Brokaw moderated that telecast. Over to the right where Suman and his family were standing for five hours by now, ABC's Dick Clark, conducting the famous *American Bandstand* and they were playing all the famous tunes including "America the Beautiful." The whole atmosphere was very festive and invigorating, as if someone was saying,

"Come, All Ye Faithful...," or as Neela would be singing, "Eso He Baisakh, Eso, Eso," actually, to Suman's utter dismay, he found Neela was singing that very song. No one could hear her, even Suman could not since the noise, chatters, laughter and loud music was deafening. Suman smiled and just said, "Baisakh in December?" She just gave a smirk.

"Baba, let's move closer to the front, the floats are just about to come, it is almost 6:00 P.M." That was almost a command from the older son which cannot be ignored by this family at any time. Suman remembers that from day one this boy has ruled their lives, of course in a very good way, even moving to Owensboro and joining the Daviess County High School in his senior year was his sole decision, rest of the family just had to oblige.

"Okay, but we will be squeezed there," little bro finally expressed his opinion.

"That's alright, you will be able to see the floats much better, and we will not be squeezed, we will manage, let's move."

American Bandstand started playing again, Dick Clark will be doing the Emcee and his assistants are getting ready; over there, where the family was squatting for the past three hours, below the MTV station, crew was harnessing gears since they will be staging live performance by the young breathtaking Disney artist Christina Aguilera, and many, and that will start around 9:00 P.M. after the floats are gone and will continue through the night with a pause for about five minutes for the BALL to DROP.

There will be about three hours of extravaganza with all kinds of floats with so many different themes and songs and dances and drums and parades of colors that the New York Sky will be gleefully welcoming once again its citizens and the citizens from around the world as it did over the past three hundred years during the past millennium.

NBC anchor Tom Brokaw's baritone voice came over the giant loudspeakers, greeting more than a million spectators from all over the country and the world and giving advance notices to the just-about-to begin marches of the bands from various parts of the country, carefully chosen after a rigorous tests and rehearsals; so many pride-bands will be marching, and then the floats will begin to parade. All are eagerly waiting, some are pretending to know who will come first, and second and on and on.

Soumit was standing with a group of Brazilian and Dutch young men and women; the way he was laughing, high-fiving them, and talking in all subjects under the sun, it seemed as if he has been friends with them forever. Little bro was closer to Suman and Neela. He was not in their age-group and still very sigh to open up. Only difference was his older brother, when they speak there is no age-difference, it seems like that. Big brother wanted him near him, however, he felt comfortable closer to his parents.

"There comes the band," someone from the nearby group yelled.

The first band was from the capital of New York, Albany. As they enter, the famous voice of Dick Clark came through the loudspeakers that were mounted selectively on the light posts all over the time square and on to the Broadway. Every time a new float appears it precedes by marching bands or a dancing party exhibiting their well-prepared acts. Suman was watching standing next to Neela, both were mesmerized by the craftsmanship and the professionalism of the performers. Suddenly, there was a huge applause, "There comes the Bhangra," the float and the dancers ahead of them are from the Indian contingent of the tristate (New York, New Jersey and Connecticut). Their float was designed as a peacock, draped in Indian tricolored flags with the Ashok Chakra on top of the peacock's head. The truck that was decorated with these giant peacock as a float had about twenty people, both men and women representing the diversity that India presents to the world. Only a very few Americans can comprehend the complexity and richness of the diverse cultural mix of India. Indian crowd from the mass gathered in this relatively warm night (about 34 degrees Fahrenheit) started echoing the song that this float was delivering, and their cheerleaders were dancing. It went on for about seven minutes, two big dance performances, one popular Bhangra (a Punjabi dance that all participants dance with their arms raised and fingers pointing to the sky and chanting popular Punjabi words, and the other was a beautiful Oddissi dance where the girls wear costumes that are designed after imaginations of various forms of animals, birds, etc.).

Time was little past nine P.M. Everyone was hungry and needed to use the Porta Potty. Food was already there since Neela managed to walk up to the restaurant near which they were sheltering for the whole day. Large pizza already sliced and separated in four brown bags, and a bottle of water. But first the bathroom.

Floats, in all, there were sixty-five of them marched and presented their talents in such a way that everyone was captivated and watched for almost three hours. It is now 11:00 P.M., the big tower clock struck 11:00 P.M. and music started from the MTV station, little to left of the family where they were standing. Everyone was anxiously awaiting for Carson Daly to announce the names of the performers and here it is: Christina Aguilera will come around 11:30 P.M., and will sing till 11:55 P.M. Crowd was going gaga. Little brother was not into pop songs back then, the older one definitely a fan of Christina and was talking to the Brazilian group with his bright big eyes all sparkling in joy. Suman noticed that he had plastic clear glass and some clear liquid is in there. He guessed it alright; it must be vodka. Soumit just turned 16 two months ago, and at this age it is a pure "No, no" to them. He never could indulge his older son into having a hard liquor at that age, even though he was pretty liberal but still cannot accept the idea. He was about to tell that to Neela, but stopped, no, this will upset her more! And the fun will be lost. Maybe later on. At times, dads need to exercise little caution to maintain family tranquility. Oh, well! he should let it go now. He looked for his younger son, found him close to his brother, about 30 feet away to the right of them, almost clinging on to his big brother, and his eyes were looking for any glass in his younger son's hand, found none and heaved a sigh of relief.

One after another singer was coming on the raised platform like a makeshift stage or a Dias at the MTV station, and now it is Christina's time to appear—she appears in a long white crystal gown and greets public. Thunderous applauds began to pour and crowd begin to cheer her up with "What a Girl Wants," her just released platinum album. She nodded and said, "Yes, I will." Sang three songs before finally sang her most famous song. Crowd went crazy, applause did not want to fade away, finally she bowed to the audience and exited.

Time was 11:53 P.M. John William's famous "Star Wars" theme songs began to play, crowd started to gather next to their kins and some were looking for their hats, blowpipes to make weird noises and some were opening plastic Champagne bottles and plastic glasses (no glass bottles were allowed in that area). Soumit was fixing their Cannon camcorder which Suman just bought for this trip. Since Soumit is the tallest (just a hair short of six feet) in the family, he has a natural claim on this new gadget, and no one asked for anything otherwise, even though Neela is the best photographer in the family. Everyone is ready. Somehow, Soumit manages to get four little clear plastic glasses, with some champagne in them, and the clock struck 11:58. Countdown begins, and everyone looking at the tower where the crystal ball is already visible. Everyone counting,

"59, 58...10, 9, 8, 7," and the ball starts to descend slowly but surely, all lit up in perfect red, white and blue dazzling lights; champagne glasses are raised, glittering hats are on and the blowpipes and the beads are all in their respective places waiting to burst out in the open full and loud like the waves are coming to the shore with massive sounds and force that will sway everyone on to the new millennium. Suman was in a trance, faintly hears—"5, 4, 3, 2, 1."

"HAPPY NEW YEAR, HAPPY NEW MILLENNIUM 2000." Fireworks started to spark the skies of Manhattan. Soumit hugs his mother first, then his brother and finally gives his father a big bear hug.

"Happy New Year, Baba!"

About the Author and the Book

Soumen N. Ghosh, Ph.D. is a professor of Economics and has been practicing the craft over thirty four years. Dr. Ghosh traveled extensively and has published numerous peer-reviewed articles and presented his research findings in international conferences all over the world. He is a product of Calcutta University and the Indian Statistical Institute. In his twenties he left India, first with a job at the United Nations with his newly-wed wife and then joined the Ph.D. program in Economics at a friendly place yet severe cold weather in Logan Utah (Utah State University). In his spare time he likes to listen to music (both Indian and Western Classical) and write stories, both fiction and non-fiction.

A Few Good Memories is a collection of short stories, mostly non-fiction that reflect his and his family's journey through time which are treasured jewels in his mind's eyes.

The first three stories have a common thread: *relation between a father and his son(s).* All the events are real and at times only embellished a little to portray a different side of Dr. Ghosh.

The book is solely dedicated to his parents and to his two sons and wife. This book could not have been completed without unflinching support from his wife, Dr. Sumita Chakraborti-Ghosh.